T0254946

Lecture Notes in Artificial Intelligence   13441

Subseries of Lecture Notes in Computer Science

Series Editors

Randy Goebel
  University of Alberta, Edmonton, Canada
Wolfgang Wahlster
  DFKI, Berlin, Germany
Zhi-Hua Zhou
  Nanjing University, Nanjing, China

Founding Editor

Jörg Siekmann
  DFKI and Saarland University, Saarbrücken, Germany

More information about this subseries at https://link.springer.com/bookseries/1244

Francisco S. Melo · Fei Fang (Eds.)

Autonomous Agents and Multiagent Systems

Best and Visionary Papers

AAMAS 2022 Workshops
Virtual Event, May 9–13, 2022
Revised Selected Papers

Editors
Francisco S. Melo (iD)
University of Lisbon
Porto Salvo, Portugal

Fei Fang (iD)
Carnegie Mellon University
Pittsburgh, PA, USA

ISSN 0302-9743   ISSN 1611-3349 (electronic)
Lecture Notes in Artificial Intelligence
ISBN 978-3-031-20178-3   ISBN 978-3-031-20179-0 (eBook)
https://doi.org/10.1007/978-3-031-20179-0

LNCS Sublibrary: SL7 – Artificial Intelligence

© Springer Nature Switzerland AG 2022, corrected publication 2024
This work is subject to copyright. All rights are reserved by the Publisher, whether the whole or part of the material is concerned, specifically the rights of translation, reprinting, reuse of illustrations, recitation, broadcasting, reproduction on microfilms or in any other physical way, and transmission or information storage and retrieval, electronic adaptation, computer software, or by similar or dissimilar methodology now known or hereafter developed.
The use of general descriptive names, registered names, trademarks, service marks, etc. in this publication does not imply, even in the absence of a specific statement, that such names are exempt from the relevant protective laws and regulations and therefore free for general use.
The publisher, the authors, and the editors are safe to assume that the advice and information in this book are believed to be true and accurate at the date of publication. Neither the publisher nor the authors or the editors give a warranty, expressed or implied, with respect to the material contained herein or for any errors or omissions that may have been made. The publisher remains neutral with regard to jurisdictional claims in published maps and institutional affiliations.

This Springer imprint is published by the registered company Springer Nature Switzerland AG
The registered company address is: Gewerbestrasse 11, 6330 Cham, Switzerland

Preface

AAMAS (the International Conference on Autonomous Agents and Multiagent Systems) is the largest and most influential conference in the area of agents and multiagent systems. It is organized annually by the International Foundation for Autonomous Agents and Multiagent Systems (IFAAMAS). The AAMAS conference series was initiated in 2002 in Bologna, Italy, as a joint event comprising the 6th International Conference on Autonomous Agents (AA), the 5th International Conference on Multiagent Systems (ICMAS), and the 9th International Workshop on Agent Theories, Architectures, and Languages (ATAL). In 2022, the 21st edition of AAMAS was held online, due to the restrictions imposed by the COVID-19 pandemic.

Besides the main program, AAMAS 2022 hosted a rich workshop program, aimed at stimulating and facilitating discussion, interaction, and comparison of approaches, methods, and ideas related to specific topics, both theoretical and applied, in the general area of autonomous agents and multiagent systems. Workshops provide an informal setting where participants have the opportunity to discuss specific technical topics in an atmosphere that fosters the active exchange of ideas and supports community development. In 2022, AAMAS hosted a total of 12 workshops.

This volume contains a selection of papers from five of these workshops:

- 13th Workshop on Optimization and Learning in Multiagent Systems (OptLearn-MAS);
- 23rd Workshop on Multi-Agent-Based Simulation (MABS);
- 6th Workshop on Agent-Based Modelling of Urban Systems (ABMUS);
- 10th Workshop on Engineering Multi-Agent Systems (EMAS); and
- 1st Workshop on Rebellion and Disobedience in AI (RaD-AI).

In particular, we include the best paper from each of the workshops. Additionally, we also include a paper deemed as visionary from MABS, ABMUS, EMAS, and RaD-AI. Given the specificities and scope of the different workshops, the nomination of both the best and the most visionary paper of each workshop relied on the corresponding organizing committee, which was charged with defining the criteria for selection and nominating the papers to be included in this volume. The criteria were as follows:

- **OptLearnMAS:** The best paper was selected based on an independent assessment by all workshop organizers in terms of relevance, originality, novelty, significance, and technical quality.
- **MABS:** All papers in MABS were reviewed by at least three members from the workshop's Program Committee, all of which were asked to assess whether the paper was a candidate for a best paper award. The workshop organizers then decided on who to nominate based on the reviewers' comments and recommendations.
- **ABMUS:** Each paper in ABMUS received at least three reviews from the workshop's Program Committee. The papers were then selected by the workshop organizers based

on the reviewers' scores and comments. The first selection criterion was the paper's overall score. In the case of a draw, the reviewer confidence and fit to ABMUST were also considered.
- **EMAS:** The papers were selected based on the reviewers' scores and comments.
- **RaD-AI:** The papers were selected based on a combination of (1) reviewer ranking, (2) category of content (i.e., visionary vs. containing concrete results), and (3) an assessment by the workshop organizers regarding the highest quality papers in the workshop in each of the two categories.

As workshop chairs, it is our belief that the AAMAS workshops provide a unique opportunity for researchers to connect and exchange ideas with other researchers working in very closely related topics and problems. We hope that the publication of this volume can help to disseminate the high-quality work that was presented and discussed in these workshops, and serve as an instrument to foster further participation in the AAMAS workshop program.

July 2022   Francisco S. Melo
                                                                                  Fei Fang

Organization

AAMAS 2022 Workshop Co-chairs

Francisco S. Melo   INESC-ID and University of Lisbon, Portugal
Fei Fang   Carnegie Mellon University, USA

AAMAS 2022 Workshop Organizers

OptLearnMAS 2022

Hau Chan   University of Nebraska-Lincoln, USA
Ferdinando Fioretto   Syracuse University, USA
Jiaoyang Li   University of Southern California, USA

MABS 2022

Fabian Lorig   Malmö University, Sweden
Emma Norling   University of Sheffield, UK

ABMUS 2022

Jason Thompson   University of Melbourne, Australia
Minh Le Kieu   University of Auckland, New Zealand
Koen van Dam   Imperial College London, UK
Nic Malleson   University of Leeds, UK
Alison Heppenstall   University of Glasgow, UK
Jiaqi Gee   University of Leeds, UK

EMAS 2022

Amit Chopra   Lancaster University, UK
Jürgen Dix   Clausthal University of Technology, Germany
Rym Zalila-Wenkstern   University of Texas at Dallas, USA

RaD-AI 2022

David W. Aha   U.S. Naval Research Laboratory, USA
Reuth Mirsky   Bar-Ilan University, Israel
Peter Stone   University of Texas at Austin, USA

Contents

Best Papers

TOPS: Transition-Based Volatility-Reduced Policy Search

Liangliang Xu[1] , Daoming Lyu[1] , Yangchen Pan[2], Aiwen Jiang[3] ,
and Bo Liu[1]

[1] Auburn University, Auburn, AL, USA
{lxz0014,daoming.lyu,boliu}@auburn.edu
[2] University of Alberta, Edmonton, AB, Canada
pan6@ualberta.ca
[3] Jiangxi Normal University, Nanchang, Jiangxi, China
jiangaiwen@jxnu.edu.cn

Abstract. Existing risk-averse reinforcement learning approaches still face several challenges, including the lack of global optimality guarantee and the necessity of learning from long-term consecutive trajectories. Long-term consecutive trajectories are prone to involving visiting hazardous states, which is a major concern in the risk-averse setting. This paper proposes **T**ransition-based v**O**latility-controlled **P**olicy **S**earch (TOPS), a novel algorithm that solves risk-averse problems by learning from transitions. We prove that our algorithm—under the overparameterized neural network regime—finds a globally optimal policy at a sublinear rate with proximal policy optimization and natural policy gradient. The convergence rate is comparable to the state-of-the-art risk-neutral policy-search methods. The algorithm is evaluated on challenging Mujoco robot simulation tasks under the mean-variance evaluation metric. Both theoretical analysis and experimental results demonstrate a state-of-the-art level of TOPS' performance among existing risk-averse policy search methods.

Keywords: Reinforcement learning · Risk control · Volatility control

1 Introduction

The world has witnessed the successes of reinforcement learning (RL, [46]) in multiple fields and domains [36]. However, there are still three concerns with existing RL approaches, which are *risk, long-term shocks*, and *global optimality*. The first concern, *risk*, refers to the instability with respect to the uncertainty of future outcomes [13], often measured by the variance of the future outcome (e.g., expected cumulative rewards). Most RL settings are risk-neutral [36,50, 53], meaning that an agent's goal is merely to learn to maximize the expected return (cumulative rewards) without considering the variance. Controlling risk is necessary in a variety of applications, including financial decision-making [27], healthcare [39], and robotics [32].

© Springer Nature Switzerland AG 2022
F. S. Melo and F. Fang (Eds.): AAMAS 2022 Workshops, LNAI 13441, pp. 3–47, 2022.
https://doi.org/10.1007/978-3-031-20179-0_1

The second concern, *long-term shocks*, is about visiting fatal or hazardous states (i.e., states with extremely low future outcomes) in the process of long-term interactions with the environment [18,20]. Unfortunately, avoiding hazardous state visitation is not always guaranteed for risk-averse RL. A key observation is that visiting hazardous states are often caused by the agent's long-term consecutive interactions with the environment [7,24,48]. Long-term consecutive interactions with the environment tend to generate trajectories with hazardous state visitations [7]. The possibility of triggering hazardous state visitations would be significantly reduced if the agent does not learn from long-term trajectories. However, most existing risk-averse RL algorithms require learning from long-term trajectories. Otherwise, one has to use an additional learning rate for the bootstrap-based critic learning, resulting in a multi-timescale step-size tuning scheme, which is quite inconvenient in practice.

The third concern regards global optimality. The theoretical understanding of policy gradient methods is under tentative study. Work on this topic has been done mostly in the tabular setting. [11] and [44] establish non-asymptotic convergence guarantees for various policy gradient methods with regularization. [35] show convergence rate for softmax parametrization. [1] analyze multiple policy gradient methods in the tabular setting as well as the linear approximation setting. [28,61] extend their work to an off-policy setting. A large spectrum of work has been done on the global optimality of policy gradient methods in a non-linear approximation setting with over-parameterized neural networks [31, 51,62].

In this paper, we aim to answer one question: *Can risk-aware policy gradient algorithms have global optimality convergence guarantee and learn safely without the need for long-term trajectories?* Motivated by addressing this question, we propose Transition-based vOlatility-controlled Policy Search (TOPS), a risk-averse RL framework with reward volatility [7] as its risk measurement and establish its global convergence and optimality. This paper makes two major contributions. *First*, instead of learning from long-term rollouts [55,57,62], our method TOPS does not require learning from long-term, uninterrupted trajectories. Instead, it can be either trained with segments of long-term rollouts, short-term trajectories, or a combination of them. This is achieved by using a lower-bound surrogate loss mean-volatility loss function (other than the original mean-variance loss function as in [14]) inspired by [7,60]. *Second*, we present a theoretical analysis of the global optimality of the proposed algorithm and prove that TOPS converges to a globally optimal policy at the rate of $1/\sqrt{K}$, where K is the number of iterations. This is achieved by the primal-dual formulation of the mean-volatility function used in [60] and the primal-dual sample complexity analysis inspired by [57,62].

The roadmap of the paper is as follows. We introduce the background in Sect. 2. In Sect. 3, we formulate the TOPS algorithm. We present the major result on its global convergence in Sect. 4. In Sect. 5, we perform experiments on benchmark domains and compare them with state-of-the-art methods. We discuss related work in Sect. 6 and conclude the paper in Sect. 7.

2   Background

This section introduces the background knowledge of the building blocks of this paper, such as reinforcement learning, policy gradient, over-parameterized neural networks. A detailed notation system is provided in Appendix A.

Reinforcement Learning. We consider the infinite-horizon discounted Markov Decision Process (MDP) $(\mathcal{S}, \mathcal{A}, \mathcal{P}, r(s, a), \gamma)$ with state space $\mathcal{S}$, action space $\mathcal{A}$, the transition kernel $\mathcal{P} : \mathcal{S} \times \mathcal{S} \times \mathcal{A} \to [0, 1]$, the reward function $r(s, a)$: $\mathcal{S} \times \mathcal{A} \to \mathbb{R}$, the initial state $S_0 \in \mathcal{S}$ and its distribution $\mu_0 : \mathcal{S} \to [0, 1]$, and the discounted factor $\gamma \to (0, 1)$. At time step t, given a state s_t, an action a_t is taken according to policy $\pi(a_t|s_t) : \mathcal{S} \times \mathcal{A} \to [0, 1]$, generating a reward $r_t := r(s_t, a_t)$ and the next state s_{t+1} based on $p(s_{t+1}|s_t, a_t)$ the reward function is assumed to be deterministic and bounded—a constant $r_{\max} > 0$ exists such that $r_{\max} = \sup_{(s,a) \in \mathcal{S} \times \mathcal{A}} |r(s, a)|$. A change in states upon an action $(s, a, r(s, a), s')$ is termed a *transition*, where the state s' is the successive state of the state s. With a little bit abuse of notation, $r(s, a)$ is denoted as $r_{s,a}$, and $r(s_t, a_t)$ is denoted as r_t in the rest of the paper. A *trajectory* of length T is a consecutive sequence of transitions $\{(s_t, a_t, r_t, s_t')\}_{t=0}^{T-1}$ over a set of contiguous timestamps, where $\forall t, s_t' = s_{t+1}$. Therefore, the trajectory is also equivalently denoted by $\{(s_t, a_t, r_t, s_{t+1})\}_{t=0}^{T-1}$. To evaluate the performance of policy π, we introduce state value function $V_\pi(s) := (1 - \gamma)\mathbb{E}_{a \sim \pi(a|s)}\left[\sum_{t=0}^{\infty} \gamma^t r_t \big| S_0 = s, a_t \sim \pi(a|s_t), s_{t+1} \sim \mathcal{P}(s|s_t, a_t)\right]$ and state-action value function $Q_\pi(s, a) := (1 - \gamma)\mathbb{E}_{a \sim \pi(a|s)}\left[\sum_{t=0}^{\infty} \gamma^t r_t \big| s_0 = s, a_0 = a, a_t \sim \pi(a|s_t), s_{t+1} \sim \mathcal{P}(s|s_t, a_t)\right]$. Bounded reward implies $|V_\pi(s)| \le r_{\max}$ and $|Q_\pi(s, a)| \le r_{\max} \; \forall \pi$. Additionally, the advantage function $A_\pi : \mathcal{S} \times \mathcal{A} \to \mathbb{R}$ of policy π is defined as $A_\pi(s, a) := Q_\pi(s, a) - V_\pi(s)$. The normalized state and state action occupancy measure of policy π is denoted by $\nu_\pi(s)$ and $\sigma_\pi(s, a) := \pi(a|s)\nu_\pi(s)$, respectively. Therefore, $\nu_\pi(s) := (1 - \gamma) \sum_{t=0}^{\infty} \gamma^t Pr(s_t = s|\mu_0, \pi, \mathcal{P})$ and $\sigma_\pi(s, a) := (1 - \gamma) \sum_{t=0}^{\infty} \gamma^t Pr(s_t = s, a_t = a|\mu_0, \pi, \mathcal{P})$, where $Pr(s_t = s|\mu_0, \pi, \mathcal{P})$ is the probability of $s_t = s$ given $\mu_0, \pi, \mathcal{P}$. Finally, the return is defined as $G := \sum_{t=0}^{\infty} \gamma^t r_t$.

Policy Gradient Methods. In the following, we discuss two policy gradient methods, where the policy π_θ is parameterized by the parameter θ. For natural policy gradient (NPG, [22]), we first define the Fisher information matrix,

$$F(\theta) := \mathbb{E}_{(s,a) \sim \sigma_{\pi_\theta}} \left[\nabla_\theta \log(\pi_\theta)(\nabla_\theta \log(\pi_\theta))^\top\right] \tag{1}$$

The update of parameter θ then takes the form,

$$\theta_{k+1} = \theta_k + \eta_{\mathrm{NPG}} \big(F(\theta_k)\big)^{-1} \nabla J_\theta(\pi_{\theta_k}) \tag{2}$$

where $\big(F(\theta_{k-1})\big)^{-1}$ is the inverse of the Fisher information matrix $F(\theta)$ in Eq. (1), ∇J_θ is the objective gradient and η_{NPG} the learning rate.

In proximal policy optimization (PPO, [43]), at the k-th iteration the update of policy parameter θ takes the following form, where β is the penalty hyper-parameter:

$$\arg\max_\theta \mathbb{E}_{(s,a) \sim \sigma_k} \left[\pi_\theta A_{\theta_k/\pi_{\theta_k}} - \beta \mathrm{KL}(\pi_\theta \| \pi_{\theta_k})\right]. \tag{3}$$

Policy Network with Over-Parameterized Neural Networks. Policy π with the two-layer over-parameterized neural network is defined as: for $\forall (s,a) \in S \times A$,

$$f((s,a); \theta, b) := \frac{1}{\sqrt{m}} \sum_{v=1}^{m} b_v \text{ReLU}((s,a)^\top [\theta]_v). \tag{4}$$

Here (s,a) is the input and m is the width of the network. $\theta = ([\theta]_1^\top, \cdots, [\theta]_m^\top)^\top \in \mathbb{R}^{m \times d}$ is the input weight matrix in the first layer of the neural network. $b = (b_1, \cdots, b_m)^\top \in \mathbb{R}^{m \times 1}$ are the output weights in the second layer. We present a block diagram of a over-parameterized neural network with Fig. 4 in the Appendix B. At the start of training, the parameters θ, b are initialized by

$$\theta = \Theta_{\text{init}} \in \mathbb{R}^{m \times d} ([\Theta_{\text{init}}]_v \sim \mathcal{N}(0, I_d/d), \tag{5}$$

and $b_v \sim \text{Unif}(\{-1,1\}), \forall v \in [m]$, respectively, where $\mathcal{N}$ denotes Gaussian distribution and Unif denotes uniform distribution. $f((s,a); \theta, b)$ can be simplified to $f((s,a); \theta)$ by updating only $[W]_v$ during training, and fixing b as its initialization [3]. We also restrict the possible value of θ within the space denoted by $\mathcal{D} = \{\xi \in \mathbb{R}^{md} : \|\xi - \Theta_{\text{init}}\|_2 \leq \Upsilon, \Upsilon > 0\}$. Therefore the policy is defined $\pi_\theta(a|s)$ in the following form, where τ is the temperature parameter.

$$\pi_\theta(a|s) := \frac{\exp(\tau f((s,a); \theta))}{\sum_{a' \in A} \exp(\tau f((s,a'); \theta))}.$$

Furthermore, the feature mapping of a two-layer neural network $f((s,a); \theta)$ is defined as, $\phi_\theta := ([\phi_\theta]_1^\top, \cdots, [\phi_\theta]_m^\top)^\top$, where $[\phi_\theta]_v^\top = \frac{b_v}{\sqrt{m}}\text{ReLU}((s,a)^\top[\theta]_v)$, $\forall v \in [m]$. By Eq. (4), it holds that $f((s,a); \theta) = \phi(s,a)^\top \theta$ and $\nabla_\theta f((s,a), \theta) = \phi(s,a)$ [51]. Furthermore, we assume that there exists a constant $M > 0$ such that,

$$\mathbb{E}_{(s,a) \sim \text{init}}\left[\sup_{(s,a) \in S \times A} |\phi((s,a)^\top \Theta_{\text{init}})|^2\right] \leq M^2.$$

Mean-Variance and Mean-Volatility RL. In a variance-constraint problem with the variance of the total reward, the objective can be formulated as,

$$\max_\pi J(\pi), \qquad \text{subject to } \mathbb{V}(G) \leq Y \tag{6}$$

where $J(\pi) := \mathbb{E}_{(s,a) \sim \sigma_\pi}[G] = \frac{1}{1-\gamma}\mathbb{E}_{(s,a) \sim \sigma_\pi}[r_{s,a}]$ is the expected return, $\mathbb{V}(\cdot)$ is the variance of a random variable and $Y > 0$ is the upper bound for this variance. The constrained formulation in Eq. (6) is NP-hard [45], and in reality, the relaxed formulation $J_\lambda^G(\pi)$ defined in Eq. (7) is often solved instead [14,26,55] as follows, where λ is called variance-controlling parameter.

$$J_\lambda^G(\pi) := \mathbb{E}[G] - \lambda \mathbb{V}(G) = \mathbb{E}[G] - \lambda \mathbb{E}[G^2] + \lambda(\mathbb{E}[G])^2 \tag{7}$$

Meanwhile, [7] proposed a reward-volatility risk measure. *Volatility* is defined as the variance of per-step reward—per-step reward R is a discrete random variable with a probability mass function of $p(R = x) = \sum_{s,a} \sigma_\pi(s,a)\mathbb{1}\{r_{s,a} = x\}$, where $\mathbb{1}\{\cdot\}$ is the indicator function. It is easy to see that $\mathbb{E}[R] = (1-\gamma)J(\pi)$ [60]. $\mathbb{V}(R)$ is the variance of R. Likewise, $J_\lambda(\pi)$ is proposed as a counterpart of Eq. (7) in the sequel, which is defined with respect to R.

$$J_\lambda(\pi) := \mathbb{E}[R] - \lambda\mathbb{V}(R) = \mathbb{E}[R] - \lambda\mathbb{E}[R^2] + \lambda(\mathbb{E}[R])^2$$

We first present the following lemma based on Lemma 1 in [7] to show that $J_\lambda(\pi)$ is a reasonable counterpart to $J_\lambda^G(\pi)$.

Lemma 1. *Given* $\lambda \geq 0$, $\frac{1}{(1-\gamma)}J_{\frac{\lambda}{(1-\gamma)}}(\pi)$ *is a lower-bound of* $J_\lambda^G(\pi)$, *i.e.,* $\frac{1}{(1-\gamma)}J_{\frac{\lambda}{(1-\gamma)}}(\pi) \leq J_\lambda^G(\pi)$.

A detailed proof is provided in Appendix D.5. Given Lemma 1, maximizing $J_\lambda^G(\pi)$ can be reduced to maximizing its lower bound $\frac{1}{(1-\gamma)}J_{\frac{\lambda}{(1-\gamma)}}(\pi)$. There are several advantages of optimizing $J_\lambda(\pi)$. Compared to optimizing $\mathbb{V}(G)$, optimizing $\mathbb{V}(R)$ is computationally easier [60]. [7] argue that $\mathbb{V}(R)$ is better at capturing short-term risk and leads to smoother trajectories that avoid possible "shocks" caused by long-horizon trajectories [7].

3 Algorithm Formulation

In this section, we present our risk-averse policy-search algorithm. In particular, we use (i) reward volatility to construct the mean-volatility objective, which circumvents the long-horizon reward issue and avoids large variance, and (ii) over-parameterized neural network [10] as the neural network architecture of the actor and the critic to facilitate global convergence analysis.

3.1 Augmented MDP

As [7] shows, reward volatility has advantages over mean-variance methods, including a smoother trajectory and much-reduced variance. Therefore, in our paper, we choose volatility as the risk measurement. Compared with the conventional mean-variance objective, the mean-volatility objective function enables the agent to learn from transitions instead of trajectories and greatly reduces the chance of getting into hazardous states due to consecutive long-horizon explorations. This can greatly help improve safety. Note that because of *compositional* expectations $(\mathbb{E}[R])^2$, double-sampling is needed, which is a heavy burden for sampling.[1]

---

[1] For more details on double-sampling and the more general compositional expectations, please refer to [30,52].

To avoid double-sampling, we resort to augmented MDP. First, note that it holds $(\mathbb{E}[R])^2 = \max_{y \in \mathbb{R}}(2\mathbb{E}[R]y - y^2)$. Then the optimization objective transforms into:

$$\max_{\pi,y} J_\lambda^y(\pi) := \mathbb{E}_{(s,a) \sim \sigma_\pi}(r_{s,a} - \lambda r_{s,a}^2 + 2\lambda r_{s,a}y) - \lambda y^2 \qquad (8)$$

We now introduce the augmented MDP, with the augmented reward defined as follows:

$$\tilde{r}_{s,a} := r_{s,a} - \lambda r_{s,a}^2 + 2\lambda r_{s,a}y \qquad (9)$$

We refer to this new MDP as the *augmented MDP* $\tilde{M} = \{\mathcal{S}, \mathcal{A}, \mathcal{P}, \tilde{r}(s,a), \gamma\}$, and denote corresponding terms by the ˜ sign—for example, the associated state value function and state-action value function are $\tilde{V}_\pi(s)$ and $\tilde{Q}_\pi(s,a)$. $\tilde{r}(s,a)$ is denoted by $\tilde{r}_{s,a}$ for notation simplicity in the remainder of this paper. We solve Eq. (8) by maximizing y and π of the augmented MDP in two steps iteratively.

3.2   Proposed Algorithms

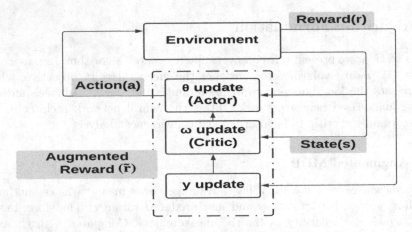

Fig. 1. A simple block diagram of TOPS

We present TOPS in Algorithm 1. A block diagram of TOPS is illustrated in Fig. 1, where there are three sets of parameters to update, i.e., θ for the actor, ω for the critic, and the auxiliary variable y. Note that the mean-volatility framework allows incorporating any off-the-shelf policy optimization methods as pointed out by [60]. Since most global optimality analysis literature is based on NPG and PPO [9,31,51,62], we also use NPG and PPO as inner policy search algorithms for a fair comparison.

y Update: Since Eq. (9) is quadratic in y, to update y in each iteration, we have $y_k = (1 - \gamma)J(\pi_k)$. However, we do not have direct access to the exact value of

$J(\pi_k)$. As an alternative, we estimate this value with a sample average in the k-th iteration,

$$\hat{y}_k := \frac{1}{T}\sum_{t=1}^{T} r_t, \tag{10}$$

as an estimator of y_k, where T is the sample (batch) size.

θ *Update:* We update π_θ with an actor-critic scheme, particularly NPG and PPO. NPG and PPO are the two most widely used policy gradient methods. According to empirical studies [54], PPO usually achieves state-of-the-art performance among on-policy policy gradient methods, and NPG has the advantage of easy hyperparameter tuning compared with PPO. On the other hand, NPG and PPO's global convergences under the risk-neutral setting have been studied intensively and can show good results [1,51]. Therefore, PPO and PNG are used as the inner actor algorithm of the TOPS framework to make fair comparisons with existing approaches. Additionally, over-parameterized neural networks are widely used in proving global convergence of gradient-based methods under the risk-neutral setting, which show impressive results [17,31,51]. The capability of a gradient-based neural network method to reach the global optimum in an over-parameterization setting is explained in theory [10]. Therefore, we parameterize the policy in the paper with a two-layer over-parameterized neural network. We first introduce the actor part of the two methods, respectively.

θ *Update for Neural NPG.* Per the update rule for NPG in Eq. (2), we need to estimate the natural policy gradient $\left(F(\theta_k)\right)^{-1}\nabla_\theta J(\pi_\theta)$. However $F(\theta_k)$ is difficult to invert due to its high-dimensionality. Instead the gradient is estimated by solving $\min_{\xi\in\mathcal{D}}\|\hat{F}(\theta_k)\xi - \tau_k\hat{\nabla}_\theta J(\pi_{\theta_k})\|_2$, where

$$\hat{\nabla}_\theta J(\pi_{\theta_k}) := \frac{\tau_k}{T}\sum_{t=1}^{T} Q_{\omega_k}(s_t, a_t)\left(\phi_{\theta_k}(s_t, a_t)\right.$$
$$\left. - \mathbb{E}_{a\sim\pi_{\theta_k}}[\phi_{\theta_k}(s_t, a'_t)]\right),$$

$$\hat{F}(\theta_k) := \frac{\tau_k^2}{T}\sum_{t=1}^{T}\left(\left(\phi_{\theta_k}(s_t, a_t) - \mathbb{E}_{a\sim\pi_{\theta_k}}[\phi_{\theta_k}(s_t, a'_t)]\right)\right.$$
$$\left.\left(\phi_{\theta_k}(s_t, a_t) - \mathbb{E}_{a\sim\pi_{\theta_k}}[\phi_{\theta_t}(s_t, a'_t)]\right)^{\top}\right),$$

are unbias estimations of $\nabla_\theta J(\pi_\theta)$ and $F(\theta_k)$ respectively, with the help of feature mapping, θ is, therefore, updated as,

$$\tau_{k+1} = \tau_k + \eta_{\mathrm{NPG}},$$
$$\theta_{k+1} = \left(\tau_k\theta_k + \eta_{\mathrm{NPG}}\arg\min_{\xi\in\mathcal{D}}\|\hat{F}(\theta_k)\xi - \tau_k\hat{\nabla}_\theta J(\pi_{\theta_k})\|_2\right)/\tau_{k+1} \tag{11}$$

θ *Update for Neural PPO.* Given the update rule of PPO in Eq. (3), the PPO's objective function $L(\theta)$ can be rewritten as $L(\theta) := \mathbb{E}_{s\sim\nu_{\pi_k}}\left[\mathbb{E}_{a\sim\pi_\theta}[Q_{\pi_k}] - \right.$

$\beta \text{KL}(\pi_\theta \| \pi_{\theta_k})$. With energy-based policy $\pi \propto \exp\{\tau^{-1}f\}$, the solution to the subproblem $\hat{\pi}_{k+1} = \arg\max_\pi L(\theta)$ can be obtained by solving the following [31]:

$$\theta_{k+1} = \arg\min_{\theta \in \mathcal{D}} \mathbb{E}_{(s,a) \sim \sigma} \left[\left(f_\theta - \tau_{k+1}(\beta^{-1}Q_{\pi_k} + \tau_k^{-1}f_{\theta_k}) \right)^2 \right] \qquad (12)$$

The stochastic gradient method can be used to solve Eq. (12).

ω *Update.* To estimate the state-action function value of the augmented MDP $\tilde{Q}_\pi$, a critic network parameterized by ω is constructed, denoted as $\tilde{Q}_\omega$. Note that the critic uses the same two-layer neural network architecture as the actor defined in Eq. (4), which indicates that the policy network π_θ's parameter θ and critic network $\tilde{Q}_\pi$'s parameter ω are of identical dimensions, i.e., $\theta \in \mathbb{R}^d, \omega \in \mathbb{R}^d$. The critic network is parameterized with a different set of parameters $\omega = ([\omega]_1^\top, \cdots, [\omega]_m^\top)^\top \in \mathbb{R}^{md}$, denoted by $f((s,a); \omega)$. For simplicity, we then learn $\tilde{Q}_\omega$ by applying the semi-gradient TD method. Other approaches, such as the Gradient TD (GTD) algorithm family [47], can also be applied. For each iteration t of the TD update,

$$\omega_{t+1} = \omega_t - \eta_{\text{TD}} \big(\tilde{Q}_{\omega_t}(s,a) \qquad (13)$$
$$- (1-\gamma)\tilde{r}_{s,a} - \gamma \tilde{Q}_{\omega_t}(s',a') \big) \nabla_\omega Q_{\omega_t}(s,a),$$

where η_{TD} is the learning rate for TD update.

4   Theoretical Analysis

Although NPG and PPO's global convergences under the risk-neutral setting show prominent result [1,51], the techniques used by these methods only apply to the primal constrained-MDP case and remain challenging to apply to the analysis of the primal-dual case as in our augmented MDP, where the dual variable y is critical. For example, Lemma (5.2) of [31], a critical step in the error-bound analysis of risk-neutral PPO, cannot be applied to our primal-dual risk-averse case. In this section, we establish the global convergence rate of TOPS with both NPG and PPO.

4.1   Assumptions

We first impose regularity condition assumptions, which are common in the literature on TD analysis with a neural network approximation [9,31,51,62].

Assumption 1 *(Variance upper bound)* [51]. *Let* $\mathcal{D} = \{\alpha \in \mathbb{R}^{md} : \|\alpha - \Theta_{\text{init}}\|_2 \leq \Upsilon\}$. *For all* $k \in [K]$, *We assume that for all* $k \in [K]$, *there exists an absolute constant* $\sigma_\xi > 0$ *such that,*

$$\mathbb{E}[\|\xi_k(\delta_k)\|_2^2] \leq \tau_k^4 \sigma_\xi^2 / T, \quad \mathbb{E}[\|\xi_k(\omega_k)\|_2^2] \leq \tau_k^4 \sigma_\xi^2 / T.$$

where $\delta_k = argmin_{\delta \in \mathcal{D}} \|\hat{F}(\theta_k)\delta - \tau_k \hat{\nabla}_\theta J(\pi_{\theta_k})\|_2$ *and* $\xi_k(\delta) = \hat{F}(\theta_k)\delta - \tau_k \hat{\nabla}_\theta \tilde{J}(\pi_{\theta_k}) - \mathbb{E}[\hat{F}(\theta_k)\delta - \tau_k \hat{\nabla}_\theta \tilde{J}(\pi_{\theta_k})]$. *The expectation is taken over* σ *given* θ_k *and* ω_k.

Algorithm 1: TOPS: **T**ransition-based V**O**latility-controlled **P**olicy **S**earch

---

1 **Input**: number of iteration K, learning rate for natural policy gradient (resp. PPO) and neural TD η_{NPG} (resp. η_{PPO}), temperature parameters $\{\tau_k\}_{k=1}^K$;

2 Initialize policy network $f((s,a); \theta, b)$ as defined in Eq. (5). Set $\tau_1 = 1$. Initialize Q-network with (b, ω_1) similarly;

3 **for** $k = 1, \cdots, K$ **do**

4 $\quad$ Sample a batch of transitions $\{s_t, a_t, r_t, s'_t\}_{t=1}^T$ following current policy with size of T;

5 $\quad$ $y = \frac{1}{T}\sum_{t=1}^T r_t$;

6 $\quad$ **for** $t = 1, \cdots, T$ **do**

7 $\quad\quad$ $\tilde{r}_t = r_t - \lambda r_t^2 + 2\lambda r_t y$, $a'_t \sim \pi(a|s'_t)$;

8 $\quad$ **end**

9 $\quad$ **Q-value update**: update ω_k according to Eq. (13);

10 $\quad$ **if** *select NPG update* **then**

11 $\quad\quad$ update θ_k according to Eq. (11) ;

12 $\quad$ **else if** *select PPO update* **then**

13 $\quad\quad$ update θ_k according to Eq. (12) ;

14 **end**

15 **Output**: π_{θ_K};

Note that δ_k and ω_k have the same dimension due to the compatible neural network setting. $\xi_k(\delta)$ can be generalized to both δ_k and ω_k. We then introduce a regularity condition assumption on visitation measures and stationary distributions in the sequel, respectively.

Assumption 2 *(Upper bounded concentrability coefficient)* [51]. *ν^* and σ^* are denoted as the state and state-action visitation measures corresponding to the global optimum π^*. For all $k \in [K]$, we define the following terms:*

$$\varphi_k = \left\{ \mathbb{E}_{(s,a)\sim\sigma_{\pi_k}}\left[\left(\frac{d\sigma^*}{d(\sigma_{\pi_k})}\right)^2\right] \right\}^{1/2}, \quad \psi_k = \left\{ \mathbb{E}_{s\sim\nu_{\pi_k}}\left[\left(\frac{d\nu^*}{d(\nu_{\pi_k})}\right)^2\right] \right\}^{1/2},$$

$$\varphi'_k = \left\{ \mathbb{E}_{(s,a)\sim\sigma'_{\pi_k}}\left[\left(\frac{d\sigma^*}{d(\sigma'_{\pi_k})}\right)^2\right] \right\}^{1/2}, \quad \psi'_k = \left\{ \mathbb{E}_{s\sim\nu'_{\pi_k}}\left[\left(\frac{d\nu^*}{d(\nu'_{\pi_k})}\right)^2\right] \right\}^{1/2}.$$

We assume that $\varphi_k, \psi_k, \varphi'_k, \psi'_k$ are uniformly upper bounded by an absolute constant $c_0 > 0$.

σ'_{π_k} and ν'_{π_k} are state-action and state stationary distribution. $\varphi_k, \psi_k, \varphi'_k, \psi'_k$ are the concentrability coefficients, which reflects how much the starting state and state-action distribution diverge from the state and state-action distribution under the optimal policy [37]. Assumption 2 impose a upper bound on such divergence. This regularity condition is commonly used in the literature [4,

16,38,51,58]. More over, we define $\varphi_k^* = \mathbb{E}_{(s,a)\sim\sigma_\pi}\left[\left(\frac{d\pi^*}{d\pi_0} - \frac{d\pi_{\theta_k}}{d\pi_0}\right)^2\right]^{1/2}, \psi_k^* = \mathbb{E}_{(s,a)\sim\sigma_\pi}\left[\left(\frac{d\sigma_{\pi^*}}{d\sigma_\pi} - \frac{d\nu_{\pi^*}}{d\nu_\pi}\right)^2\right]^{1/2}.$

4.2 Major Theoretical Results

In the following, we present the major theoretical results, i.e., the global optimality and convergence rate of TOPS with neural PPO. We define the optimality gap $\min_{k\in[K]}\left(J_\lambda^{y^*}(\pi^*) - J_\lambda^{\hat{y}_k}(\pi_k)\right)$, where π^*, y^* are respectively defined as $\pi^* := \arg\max_\pi J_\lambda(\pi)$, $\hat{y}^* := (1-\gamma)J(\pi^*)$, and y^k is defined in Eq. (10). $J_\lambda^{y^*}(\pi^*)$ (resp. $J_\lambda^{\hat{y}_k}(\pi_k)$) represents the risk-averse objective under π^* (resp. π_k, i.e., the policy at the k-th iteration).

Theorem 1 *(Global Optimality and Rate of Convergence on neural PPO). We set the learning rate of PPO $\eta_{\text{PPO}} = \min\{(1-\gamma)/3(1+\gamma)^2, 1/\sqrt{K_{\text{TD}}}\}$, the learning rate of TD update $\eta_{\text{TD}} = \min\{(1-\gamma)/3(1+\gamma)^2, 1/\sqrt{K_{\text{TD}}}\}$ where K_{TD} is the total iteration of TD update, and $\beta_0 := \beta/\sqrt{K}$. Under Assumptions 3–4, we have, with a probability of $1-\delta$,*

$$\min_{k\in[K]}\left(J_\lambda^{y^*}(\pi^*) - J_\lambda^{\hat{y}_k}(\pi_k)\right)$$
$$\leq \frac{\beta_0^2\log|\mathcal{A}| + U + \beta_0^2\sum_{k=1}^K(\varepsilon_k + \varepsilon_k')}{(1-\gamma)\beta_0\sqrt{K}}$$
$$+ \frac{4\lambda c_3 r_{\max}(1-\gamma)}{\sqrt{K}}$$

where $\varepsilon_k'' = \mathcal{O}(\Upsilon^3 m^{-1/2}\log(1/\delta) + \Upsilon^{5/2}m^{-1/4}.$
$$\sqrt{\log(1/\delta)} + \Upsilon\cdot r_{\max}^2 m^{-1/4} + \Upsilon^2 K_{\text{TD}}^{-1/2} + \Upsilon),$$
$$\varepsilon_k = \tau_{k+1}^{-1}\varepsilon_k''\varphi_{k+1}^* + \beta^{-1}\varepsilon_k''\psi_k^*,$$
$$\varepsilon_k' = |\mathcal{A}|\tau_{k+1}^{-2}\epsilon_{k+1}^2,$$
$$U = 2\mathbb{E}_{s\sim\nu_{\pi^*}}[\max_{a\in\mathcal{A}}(\tilde{Q}_{\omega_0})^2] + 2\Upsilon^2.$$

Similarly, we present TOPS global optimality and convergence rate with neural NPG.

Theorem 2 *(Global Optimality and Rate of Convergence for neural NPG). We set the learning rate of NPG $\eta_{\text{NPG}} = 1/\sqrt{K}$, the learning rate of TD update $\eta_{\text{TD}} = \min\{(1-\gamma)/3(1+\gamma)^2, 1/\sqrt{K_{\text{TD}}}\}$ where K_{TD} is the total iteration of TD update, and the temperature parameters $\tau_k = (k-1)\eta_{\text{NPG}}$. Under Assumptions 3–4, with a probability of $1-\delta$, we have*

$$\min_{k\in[K]}\left(J_\lambda^{y^*}(\pi^*) - J_\lambda^{\hat{y}_k}(\pi_k)\right)$$

$$\leq \frac{1}{(1-\gamma)\sqrt{K}}\left(\log|\mathcal{A}| + 9\Upsilon^2 + M^2 + 4c_3 M(1-\gamma)^2\lambda\right)$$

$$+ \frac{1}{K}\sum_{k=1}^{K}(\epsilon_k)$$

$$where \ \epsilon_k = \sqrt{8}c_0\Upsilon^{1/2}\sigma_\xi^{1/2}T^{-1/4}$$

$$+ \mathcal{O}\big((\tau_{k+1}K^{1/2}+1)\Upsilon^{3/2}m^{-1/4} + \Upsilon^{5/4}m^{-1/8}\big)$$

$$+ c_0\mathcal{O}(\Upsilon^3 m^{-1/2}\log(1/\delta) + \Upsilon^{5/2}m^{-1/4}\sqrt{\log(1/\delta)}$$

$$+ \Upsilon r_{\max}^2 m^{-1/4} + \Upsilon^2 K_{\mathrm{TD}}^{-1/2} + \Upsilon)$$

Remark 1. Theorem 1 and 2 show the upper bound of the optimality gap

$$\min_{k\in[K]}\left(J_\lambda^{y^*}(\pi^*) - J_\lambda^{\hat{y}_k}(\pi_k)\right) \sim O(\frac{1}{K}),$$

where K is the maximum number of updates. It reflects how close the policy produced by TOPS can achieve to the global optimal policy.

From Theorem 1 and 2, we can conclude that our risk-averse algorithm TOPS with PPO and NPG version of the actor both converge to the global optimal policy at a $\mathcal{O}(1/\sqrt{K})$ rate. We provide a detailed proof in Appendix D.

5   Experiments

In this section, we aim to empirically examine the performance of our algorithm on Mujoco robot manipulation benchmark tasks from OpenAI gym [8], as [60] does. The Mujoco benchmark is a set of challenging robot control tasks in simulated environments designed for controller optimization in reinforcement learning [49]. In this domain, the simulated robots are expected to achieve consistent performances while avoiding failures that lead to dangerous results.

Experiment Setup. We conduct our experiments in an online learning setting and include several recent risk-averse RL methods as baselines: the mean-variance policy optimization (MVP) [55], mean-variance policy iteration (MVPI) [60], and variance-constrained actor-critic (VARAC) [62]. We set $\lambda = 1$ and run each algorithm for 10^6 steps and evaluate the algorithm every 10^4 steps for 20 episodes. All curves are averaged over 10 independent runs and use shaded areas to indicate standard errors. The experiment's details are provided in Appendix C. All experiments' parameters are tuned through rigorous grid search.

We report the learning curves of TOPS with NPG and PPO, respectively, in Fig. 2 and 3. For MVP, since the algorithm uses coordinate gradient at each step, it does not have a PPO or NPG version, and therefore we only report its learning curve in Fig. 2. For MVPI, we use its on-policy version for the experiment. As it works with any off-the-shelf policy search method, we implement NPG and

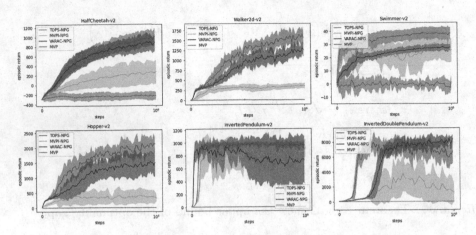

Fig. 2. Training progress of TOPS-NPG and baseline algorithms.

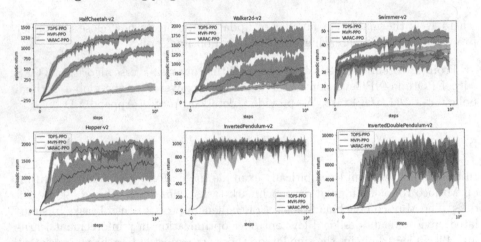

Fig. 3. Training progress of TOPS-PPO and baseline algorithms.

PPO with two-layer over-parameterized neural networks as its policy search component.

There are several interesting findings of our experiments. *First*, the results show that TOPS outperforms other baselines in most of the testbeds with respect to initial learning speed, the variance during the learning process, and the steady-state mean. In particular, TOPS outperforms other methods with a large margin with respect to the mean and variance of the learning curve on `Walker2d-v2`, `Hopper-v2` and `InvertedPendulum-v2`, as seen from Fig. 2 and Fig. 3. *Second*, the partial order of performance level tends to be consistent across different tasks, regardless of the base method (NPG or PPO). The order of performance level (from the best to the worst) in most subfigures of Fig. 2 is TOPS-NPG, VARAC-NPG, and MVPI-NPG. Similar results are observed in the majority of the subfigures in Fig. 3, where the order is TOPS-PPO, VARAC-PPO, and MVPI-PPO. Compared to other methods, MVP performs poorly in all tested domains, which indicates it may not suit the tasks. Note that on domains such as Walker2d-v2 and Hopper-v2, MVP's curves show zero variance with zero mean. This is due to the fact that these domains have a sparse reward nature (i.e., the reward is 0 for most states), and zero mean and zero reward indicate that MVP simply learns nothing useful. Overall, these results demonstrate that our TOPS algorithm can achieve state-of-the-art risk-averse performance on the challenging robot simulator testbeds.

6   Related Work

In risk-averse RL, variance is a more popular risk measure among its many peers [14,26,33,45,55], with the related approaches usually referred to as mean-variance RL. Variance stands out due to its advantage in interpretability and computation [29,34]. Most mean-variance RL methods consider the variance of the total reward [14,26,55]. In contrast, [7] and [60] propose a reward-volatility risk measure using the variance of a per-step reward. They show that the reward-volatility method is better at capturing the short-term risk and easier to compute.

In particular, we distinguish our method with [7,55], and [60]. [55] utilize the variance of the cumulative rewards. However, this method's theoretical analysis is limited to the sample complexity of episodic average-reward MDP, and the choice of solvers is restricted. Both [7] and [60] use the variance of the per-step reward, which introduces a policy-dependent-reward issue. [7] solve this directly, but it is much more difficult than normal MDP due to the lack of tools, and this approach also requires double-sampling. [60] avoid these issues by proposing an augmented MDP in a flexible framework that can apply any off-the-shelf policy evaluation and control method and name it MVPI. Our paper is inspired by [60], but our approach differs from MVPI. At any iteration, MVPI needs to keep updating π until it maximizes the objective function defined in Eq. (8) with a fixed y, while TOPS is only required to update π once. This key difference enables TOPS to train faster, as shown in Sect. 5. We present an algorithm comparison between TOPS and MVPI in Algorithm 2 of the Appendix B. Furthermore, we can also provide a theoretical analysis of global convergence.

Second, we compare our proof technique with those of [31,51,57,62]. These papers share a similar method in the first part of the analysis, but the second parts are different because each works on a different policy gradient method. Our paper adopts the methods of [51] and [31] for neural NPG and neural PPO, respectively, and develops a method for the Q-function value based on the two. Compared with [51] and [31], our results of convergence are in probability rather than expectation because we utilize different techniques when we characterize certain error bounds. Specifically, we develop a high probability bound for the critic part while using an expectation bound in the actor part. The closest work to ours is [62], which utilizes the variance of the cumulative rewards and therefore needs to learn from consecutive trajectories instead of non-consecutive transitions. A corresponding disadvantage is that it requires two critics to represent value functions associated with the original reward and the squared reward. On the contrary, TOPS only needs one due to the deployment of the augmented MDP. Additionally, while they present theoretical proofs, they do not provide empirical results. Moreover, in their proof, Eq. (4.15) does not hold, which leads to an invalid core conclusion regarding their Eq. (4.17), an essential part of the proof. Therefore, the validity of the theoretical analysis starting from their Eq. (4.15) remains unclear. The details are mentioned in Appendix D.2. The most recent work along this research line is [57]. Unlike other works, they solve safe reinforcement learning problems in the primal space, termed CRPO. In addition, the theoretical analysis of CRPO is restricted to a simplified version of NPG. In contrast, we present theoretical proof for neural NPG and PPO as the policy search algorithms. Other related work includes [56,63], and [6]. A more detailed discussion is provided in Appendix E.

7 Conclusion

This paper aims to answer the following question: *can risk-averse algorithms have a global convergence guarantee and learn from short trajectories?* Theoretical analysis for both neural NPG and neural PPO with two-layer over-parameterized neural networks are presented to show that TOPS can find the global optimality at an $\mathcal{O}(1/\sqrt{K})$ converge rate. We also demonstrate the empirical success of TOPS in Mujoco robot simulation domains.

Acknowledgment. BL's research is funded by the National Science Foundation (NSF) under grant NSF IIS1910794, Amazon Research Award, and Adobe gift fund.

A   Notation Systems

- $(\mathcal{S}, \mathcal{A}, \mathcal{P}, r, \gamma)$ with state space $\mathcal{S}$, action space $\mathcal{A}$, the transition kernel $\mathcal{P}$, the reward function r, the initial state S_0 and its distribution μ_0, and the discounted factor γ.
- $r_{\max} > 0$ is a constant as the upper bound of the reward.
- State value function $V_\pi(s)$ and state-action value function $Q_\pi(s, a)$.
- The normalized state and state action occupancy measure of policy π is denoted by $\nu_\pi(s)$ and $\sigma_\pi(s, a)$
- T is the length of a *trajectory*.
- The return is defined as G. $J(\pi)$ is the expectation of G.
- Policy π_θ is parameterized by the parameter θ.
- τ is the temperature parameter in the softmax parameterization of the policy.
- $F(\theta)$ is the Fisher information matrix.
- $\eta_T D$ is the learning rate of TD update. Similarly, $\eta_N PG$ is the learning rate of NPG update. $\eta_P PO$ is the learning rate of PPO update.
- β is the penalty factor of KL difference in PPO update.

- $f\big((s,a);\theta\big)$ is the two-layer over-parameterized neural network, with m as its width.
- ϕ_θ is the feature mapping of the neural network.
- $\mathcal{D}$ is the parameter space for θ, with Υ as its radius.
- $M > 0$ is a constant as the initialization upper bound on θ.
- $J_\lambda^G(\pi)$ is the mean-variance objective function.
- $J_\lambda(\pi)$ is the reward-volatility objective function, with λ as the penalty factor.
- $J_\lambda^y(\pi)$ is the transformed reward-volatility objective function, with y as the auxiliary variable.
- $\tilde{r}$ is the reward for the augmented MDP. Similarly, $\tilde{V}_\pi(s)$ and $\tilde{Q}_\pi(s,a)$ are state value function and state-action value function of the augmented MDP, respectively. $\tilde{J}(\pi)$ is the risk-neural objective of the augmented MDP.
- $\hat{y}_k$ is an estimator of y at k-th iteration.
- ω is the parameter of critic network.
- $\delta_k = \operatorname{argmin}_{\delta \in \mathcal{D}} \|\hat{F}(\theta_k)\delta - \tau_k \hat{\nabla}_\theta J(\pi_{\theta_k})\|_2$.
- $\xi_k(\delta) = \hat{F}(\theta_k)\delta - \tau_k \hat{\nabla}_\theta \tilde{J}(\pi_{\theta_k}) - \mathbb{E}[\hat{F}(\theta_k)\delta - \tau_k \hat{\nabla}_\theta \tilde{J}(\pi_{\theta_k})]$.
- σ_ξ is a constant associated with the upper bound of the gradient variance.
- $\varphi_k, \psi_k, \varphi_k', \psi_k'$ are the concentability coefficients, upper bounded by a constant $c_0 > 0$.
- $\varphi_k^* = \mathbb{E}_{(s,a)\sim\sigma_\pi}\left[\left(\dfrac{d\pi^*}{d\pi_0} - \dfrac{d\pi_{\theta_k}}{d\pi_0}\right)^2\right]^{1/2}$.
- $\psi_k^* = \mathbb{E}_{(s,a)\sim\sigma_\pi}\left[\left(\dfrac{d\sigma_{\pi^*}}{d\sigma_\pi} - \dfrac{d\nu_{\pi^*}}{d\nu_\pi}\right)^2\right]^{1/2}$.
- K is the total number of iterations. Similarly, K_{TD} is the total number of TD iterations.
- $c_3 > 0$ is a constant as to quantify the difference in risk-neutral objective between optimal policy and any policy.

Algorithm 2: A comparison between TOPS and MVPI

---

1 **for** $k = 1, \ldots, K$ **do**
2   **Step 1:** $y_k := (1-\gamma)J(\pi_k)$;
3   **Step 2:** $\tilde{J}(\pi_{\theta_k}) := \mathbb{E}_{(s,a)\sim\sigma_{\pi_\theta}}(r_{s,a} - \lambda r_{s,a}^2 + 2\lambda r_{s,a}y_k)$;
4   **if** *MVPI:* **then**
5         $\theta_k := \arg\max_\theta(\tilde{J}(\pi_{\theta_k}))$;
        `// This is achieved by line 9 to 15 in Algorithm 3`
6   **else if** *TOPS:* **then**
7   **if** *select NPG update* **then**
8   update θ_k according to Eq. (11) ;
9   **else if** *select PPO update* **then**
10   update θ_k according to Eq. (12) ;
11 **end**
12 **Output:** π_{θ_K};

B   Algorithm Details

We provide a comparison between MVPI and TOPS. Note that neither NPG nor PPO solve $\theta_k := \arg\max_\theta(\tilde{J}(\pi_{\theta_k}))$ directly, but instead solve an approximation optimization problem at each iteration. We provide pseudo-code for the implementation of MVPI and VARAC in Algorithm 3 and 4.

C   Experimental Details

Note that although the mean-volatility method can be adapted to off-policy methods [60], in this paper, for the ease of the theoretical analysis, our proposed method is an on-policy actor-critic algorithm.

C.1   Testbeds

We use six Mujoco tasks from Open AI gym [8] as testbeds. They are Half Cheetah-v2, Hopper-V2, Swimmer-V2, Walker2d-V2, InvertedPendulum-v2, and InvertedDoublePendulum-v2.

C.2   Hyper-parameter Settings

In the experiment we set $\lambda = 1$. We then tune learning rate for different algorithms. For MVP, we use the same setting as [60]. For MVPI, TOPS and VARAC with neural NPG, we tune the learning rate of the actor network from $\{0.1, 1 \times 10^{-2}, 1 \times 10^{-3}, 7 \times 10^{-4}\}$ and the learning rate of the critic network from $\{1 \times 10^{-2}, 1 \times 10^{-3}, 7 \times 10^{-4}\}$. For MVPI, TOPS and VARAC with neural PPO, we tune the learning rate of the actor network from $\{3 \times 10^{-3}, 3 \times 10^{-4}, 3 \times 10^{-5}\}$ and the learning rate of the critic network from $\{1 \times 10^{-2}, 1 \times 10^{-3}, 1 \times 10^{-4}\}$.

Algorithm 3: MVPI with over-parameterized networks

---

1 **Input**: number of iteration K, learning rate for natural policy gradient (resp. PPO) TD η_{NPG} (resp. η_{PPO}), temperature parameters $\{\tau_k\}_{k=1}^{K}$;

2 **Initialization: Initialization**: Initialize policy network $f((s,a);\theta,b)$ as defined in Eq. (5). Set $\tau_1 = 1$. Initialize Q-network with (b,ω_1) similarly;

3 **for** $k = 1, \cdots, K$ **do**

4   Sample a batch of transitions $\{s_t, a_t, r_t, s_t'\}_{t=1}^{T}$ following current policy with size of T;

5      $y = \frac{1}{T}\sum_{t=1}^{T} r_t$;

6   **for** $t = 1, \cdots, T$ **do**

7          $\tilde{r}_t = r_t - \lambda r_t^2 + 2\lambda r_t y$, $a_t' \sim \pi(a|s_t')$;

8   **end**

9   **repeat**

10   **Q-value update**: update ω_k according to Eq. (13);

11   **if** *select NPG update* **then**

12   update θ_k according to Eq. (11);

13   **else if** *select PPO update* **then**

14   update θ_k according to Eq. (12);

15   **until** *CONVERGE*;

16 **end**

17 **Output**: π_{θ_K};

C.3 Computing Infrastructure

We conducted our experiments on a GPU GTX 970 and GPU GTX 1080Ti.

D Theoretical Analysis Details

In this section, we discuss the theoretical analysis in detail. We first present the overview in Sect. D.1. Then we provide additional assumptions in Sect. D.2. In the rest of the section, we present all the supporting lemmas and the proof for Theorem 1 and 2.

D.1 Overview

We provide Fig. 5 to illustrate the structure of the theoretical analysis. First, under Assumption 3 and 4, as well as Lemma 13. We can obtain Lemma 14, 15 and 16. These are the building blocks of Lemma 2, which is a shared component in the analysis of both NPG and PPO. The shared components also include Lemma 3, as well as Lemma 4 obtained under Assumption 5. For PPO analysis, under Assumption 2 and 4, we obtain Lemma 7 and 8 from Lemma 2 and 6, Then combined with Lemma 3, 4 and 9, we obtain Theorem 1, the major result of PPO analysis. Likely for NPG analysis, we first obtain Lemma 11 and 12 under Assumption 1, 2 and 4. Then together with Lemma 2, 3, 4 and 10, we obtain Theorem 2, the major result of NPG analysis.

Algorithm 4: VARAC

---

1 **Input**: number of iteration K, learning rate for natural policy gradient (resp. PPO) TD η_{NPG} (resp. η_{PPO}), temperature parameters $\{\tau_k\}_{k=1}^{K}$;

2 **Initialization**: Initialize policy network $f((s,a);\theta,b)$ as defined in Eq. (5). Set $\tau_1 = 1$. Initialize Q-network with (b,ω_1) similarly;

3 **for** $k = 1, \cdots, K$ **do**

4   Sample a batch of transitions $\{s_t, a_t, r_t, s_t'\}_{t=1}^{T}$ following current policy with size of T;

5      $y = \frac{1}{T} \sum_{t=1}^{T} r_t$;

6   **Q-value update**: update both networks' ω_k according to Eq. (13);

7   Output Q_k and W_k;

8   update θ_k with NPG or PPO;

9 **end**

10 **Output**: π_{θ_K};

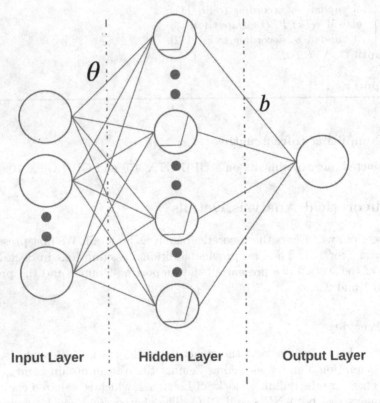

Fig. 4. A block diagram of over-parameterized neural network

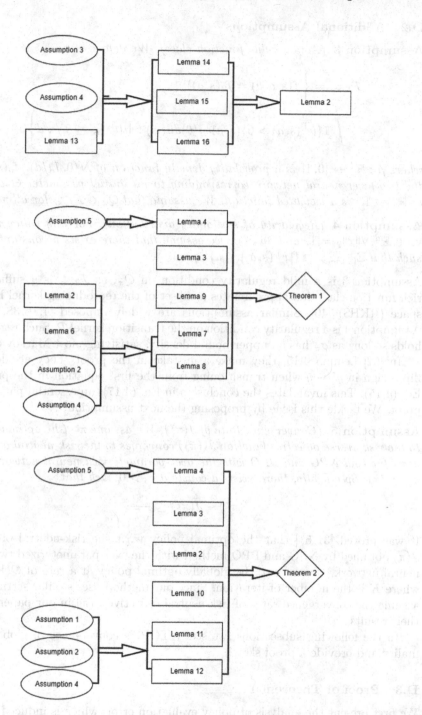

Fig. 5. A flow chart of the theoretical analysis

D.2   Additional Assumptions

Assumption 3 *(Action-value function class). We define*

$$\mathcal{F}_{\Upsilon,\infty} := \left\{ f(s,a;\theta) = f_0(s,a) \right.$$

$$\left. + \int \mathbb{1}\{\theta^\top(s,a) > 0\}(s,a)^\top \iota(\theta)d\mu(w) : \|\iota(\theta)\|_\infty \le \Upsilon/\sqrt{d} \right\}$$

where $\mu : \mathbb{R}^d \to [0,1]$ is a probability density function of $\mathcal{N}(0, I_d/d)$. $f_0(s,a)$ is the two-layer neural network corresponding to the initial parameter Θ_{init}, and $\iota : \mathbb{R}^d \to \mathbb{R}^d$ is a weighted function. We assume that $\tilde{Q}_\pi \in \mathcal{F}_{\Upsilon,\infty}$ for all π.

Assumption 4 *(Regularity of stationary distribution). For any policy π, and $\forall x \in \mathbb{R}^d, \forall \|x\|_2 = 1$, and $\forall u > 0$, we assume that there exists a constant $c > 0$ such that $\mathbb{E}_{(s,a)\sim\sigma_\pi}\left[\mathbb{1}\{|x^\top(s,a)| \le u\}\right] \le cu$.*

Assumption 3 is a mild regularity condition on Q_π, as $\mathcal{F}_{\Upsilon,\infty}$ is a sufficiently rich function class and approximates a subset of the reproducing kernel Hilbert space (RKHS) [40]. Similar assumptions are widely imposed [4,16,38,51,58]. Assumption 4 is a regularity condition on the transition kernel $\mathcal{P}$. Such regularity holds so long as σ_π has an upper bound density, satisfying most Markov chains.

In [62] Lemma 4.15, they make a mistake in the proof. They accidentally flip a sign in $y^* - \bar{y}$ when transitioning from the first equation in the proof to Eq. (4.15). This invalidates the conclusion in Eq. (4.17), an essential part of the proof. We tackle this issue by proposing the next assumption.

Assumption 5 *(Convergence Rate of $J(\pi)$). We assume π^* (the optimal policy to the risk-averse objective function $J_\lambda(\pi)$) converges to the risk-neutral objective $J(\pi)$ for both NPG and PPO with the over-parameterized neural network to be $\mathcal{O}(1/\sqrt{k})$. Specifically, there exists a constant $c_3 > 0$ such that,*

$$J(\pi^*) - J(\pi_k) \le \frac{c_3}{\sqrt{k}}$$

It was proved [31,51] that the optimal policy w.r.t the risk-neutral objective $J(\pi)$ obtained by NPG and PPO method with the over-parameterized two-layer neural network converges to the globally optimal policy at a rate of $\mathcal{O}(1/\sqrt{K})$, where K is the number of iteration. Since our method uses similar settings, we assume the convergence rates of risk-neutral objective $J(\pi)$ in our paper follow their results.

In the following subsections, we study TOPS's convergence of global optimality and provide a proof sketch.

D.3   Proof of Theorem 1

We first present the analysis of policy evaluation error, which is induced by TD update in Line 9 of Algorithm 1. We characterize the policy evaluation error in the following lemma:

Lemma 2 *(Policy Evaluation Error). We set learning rate of TD $\eta_{TD} = \min\{(1-\gamma)/3(1+\gamma)^2, 1/\sqrt{K_{TD}}\}$. Under Assumption 3 and 4, it holds that, with probability of $1-\delta$,*

$$\|\tilde{Q}_{\omega_k} - \tilde{Q}_{\pi_k}\|^2_{\nu_{\pi_k}}$$
$$= \mathcal{O}(\Upsilon^3 m^{-1/2}\log(1/\delta) + \Upsilon^{5/2}m^{-1/4}\sqrt{\log(1/\delta)}$$
$$+ \Upsilon r_{\max}^2 m^{-1/4} + \Upsilon^2 K_{TD}^{-1/2} + \Upsilon), \tag{14}$$

where $\tilde{Q}_{\pi_k}$ is the Q-value function of the augmented MDP, and $\tilde{Q}_{\omega_k}$ is its estimator at the k-th iteration. We provide the proof and its supporting lemmas in Appendix D.6. In the following, we establish the error induced by the policy update. Equation (8) can be re-expressed as

$$J_\lambda^y(\pi) = \sum_{s,a} \sigma_\pi\left(r_{s,a} - \lambda r_{s,a}^2 + 2\lambda r_{s,a}y_{k+1}\right) - \lambda y_{k+1}^2 \tag{15}$$

It can be shown that $\forall\pi, \max_y J_\lambda^y(\pi) = J_\lambda(\pi)$ [55,60]. We denote the optimal policy to the augmented MDP associated with y^* by $\pi^*(y^*)$. By definition, it is obvious that π^* and $\pi^*(y^*)$ are equivalent. For simplicity, we will use the unified term π^* in the rest of the paper. We present Lemma 3 and 4.

Lemma 3 *(Policy's Performance Difference). For mean-volatility objective w.r.t. auxiliary variable y as $J_\lambda^y(\pi)$ defined in Eq. (15). For any policy π and π', we have the following,*

$$J_\lambda^y(\pi') - J_\lambda^y(\pi) = (1-\gamma)^{-1}\mathbb{E}_{s\sim\nu_{\pi'}}\left[\mathbb{E}_{a\sim\pi'}[\tilde{Q}_{\pi,y}]\right.$$
$$\left. - \mathbb{E}_{a\sim\pi}[\tilde{Q}_{\pi,y}]\right],$$

where $\tilde{Q}_{\pi,y}$ is the state-action value function of the augmented MDP, and its rewards are associated with y.

Proof. When y is fixed,

$$J_\lambda^y(\pi') - J_\lambda^y(\pi)$$
$$= \sum_{s,a} \sigma_{\pi'}\tilde{r}_{s,a} - \sum_{s,a}\sigma_\pi\tilde{r}_{s,a} = \tilde{J}(\pi')) - \tilde{J}(\pi) \tag{16}$$

We then follow Lemma 6.1 in [21]:

$$\tilde{J}(\pi') - \tilde{J}(\pi) = (1-\gamma)^{-1}\mathbb{E}_{(s,a)\sim\sigma_{\pi'}}\left[\tilde{A}_\pi\right] \tag{17}$$

where $\tilde{A}_\pi = \tilde{Q}_\pi - \tilde{V}_\pi$ is the advantage function of policy π. Meanwhile,

$$\mathbb{E}_{a\sim\pi'}[\tilde{A}_\pi] = \mathbb{E}_{a\sim\pi'}[\tilde{Q}_\pi] - \tilde{V}_\pi = \mathbb{E}_{a\sim\pi'}[\tilde{Q}_\pi] - \mathbb{E}_{a\sim\pi}[\tilde{Q}_\pi] \tag{18}$$

From Eq. (16), Eq. (17) and Eq. (18), we complete the proof.

Lemma 3 is inspired by [21] and adopted by most work on global convergence [1, 31,57]. Next, we derive an upper bound for the error of the critic update in Line 5 of Algorithm 1:

Lemma 4 *(y Update Error). We characterize the error induced by the estimation of auxiliary variable y w.r.t the optimal value y^* at k-th iteration as, $J_\lambda^{y^*}(\pi^*) - J_\lambda^{\hat{y}_k}(\pi^*) = \frac{2c_3 r_{max}(1-\gamma)\lambda}{\sqrt{k}}$, where r_{max} is the bound of the original reward, and c_3 is a constant error term.*

Proof. We start from the subproblem objective defined in Eq. (15) with y^* and $\hat{y}_k$:

$$
J_\lambda^{y^*}(\pi^*) - J_\lambda^{\hat{y}_k}(\pi^*)
$$

$$
= \left(\sum_{s,a} \sigma_{\pi^*}(r_{s,a} - \lambda r_{s,a}^2 + 2\lambda r_{s,a} y^*) - \lambda y^{*2} \right)
$$

$$
- \left(\sum_{s,a} \sigma_{\pi^*}(r_{s,a} - \lambda r_{s,a}^2 + 2\lambda r_{s,a} \hat{y}_k) - \lambda \hat{y}_k^2 \right)
$$

$$
= 2\lambda \left(\sum_{s,a} \sigma_{\pi^*} r_{s,a} \right)(y^* - \hat{y}_k) - \lambda(y^{*2} - \hat{y}_k^2)
$$

$$
= \lambda \langle y^* - \hat{y}_k, 2(1-\gamma)J(\pi^*) - y^* - \hat{y}_k \rangle
$$

$$
= (1-\gamma)\lambda \langle y^* - \hat{y}_k, J(\pi^*) - \hat{J}(\pi_k) \rangle
$$

where we obtain the final two equalities by the definition of J_π and y. Because $r_{s,a}$ is upper-bounded by a constant r_{max}, we have $|y^* - \hat{y}_k| \leq 2r_{max}$. Under Assumption 5 we have,

$$
J_\lambda^{y^*}(\pi^*) - J_\lambda^{\hat{y}_k}(\pi^*) = \frac{2c_3 r_{max}(1-\gamma)\lambda}{\sqrt{k}}
$$

Thus we finish the proof.

From Lemma 3 and 4, we can also obtain the following Lemma.

Lemma 5 *(Performance Difference on π and y). For mean-volatility objective w.r.t. auxiliary variable y as $J_\lambda^y(\pi)$ defined in Eq. (15). For any π, y and the optimal $\pi*, y*$, we have the following,*

$$
J_\lambda^{y^*}(\pi^*) - J_\lambda^y(\pi) = (1-\gamma)^{-1} \mathbb{E}_{s \sim \nu_{\pi^*}} \left[\mathbb{E}_{a \sim \pi^*}[\tilde{Q}_{\pi,y}] \right.
$$

$$
\left. - \mathbb{E}_{a \sim \pi}[\tilde{Q}_{\pi,y}] \right] + \frac{2c_3 r_{max}(1-\gamma)\lambda}{\sqrt{k}}.
$$

where $\tilde{Q}_{\pi,y}$ is the state-action value function of the augmented MDP, and its rewards are associated with y.

Proof. It is easy to see that $J_\lambda^{y^*}(\pi^*) - J_\lambda^y(\pi) = J_\lambda^{y^*}(\pi^*) - J_\lambda^y(\pi^*) + J_\lambda^y(\pi^*) - J_\lambda^y(\pi)$. Then replace $J_\lambda^y(\pi^*) - J_\lambda^y(\pi)$ with Lemma 3 and $J_\lambda^{y^*}(\pi^*) - J_\lambda^y(\pi^*)$ with Lemma 4, we finish the proof.

Lemma 5 quantifies the performance difference of $J_\lambda^y(\pi)$ between any pair π, y and the optimal $\pi*, y*$, while Lemma 3 only quantifies the performance difference of $J_\lambda^y(\pi)$ between π and π' when y is fixed.

We now study the global convergence of TOPS with neural PPO as the policy update component. First, we define the neural PPO update rule.

Lemma 6 [31]. *Let $\pi_{\theta_k} \propto \exp\{\tau_k^{-1} f_{\theta_k}\}$ be an energy-based policy. We define the update*

$$\hat{\pi}_{k+1} = \arg\max_\pi \mathbb{E}_{s \sim \nu_k}[\mathbb{E}_\pi[Q_{\omega_k}] - \beta_k KL(\pi_\theta \| \pi_{\theta_k})],$$

where Q_{ω_k} is the estimator of the exact action-value function $Q^{\pi_{\theta_k}}$. We have

$$\hat{\pi}_{k+1} \propto \exp\{\beta_k^{-1} Q_{\omega_k} + \tau_k^{-1} f_{\theta_k}\}$$

And to represent $\hat{\pi}_{k+1}$ with $\pi_{\theta_{k+1}} \propto \exp\{\tau_{k+1}^{-1} f_{\theta_{k+1}}\}$, we solve the following sub-problem,

$$\theta_{k+1} = \arg\min_{\theta \in \mathbb{D}} \mathbb{E}_{(s,a) \sim \sigma_k}[(f_\theta(s,a) - \tau_{k+1}(\beta_k^{-1} Q_{\omega_k}(s,a)$$
$$+ \tau_k^{-1} f_{\theta_k}(s,a)))^2]$$

We analyze the policy improvement error in Line 13 of Algorithm 1. [31] proves that the policy improvement error can be characterized similarly to the policy evaluation error as in Eq. (14). Recall $\tilde{Q}_{\omega_k}$ is the estimator of Q-value, f_{θ_k} the energy function for policy, and $f_{\hat{\theta}}$ its estimator. We characterize the policy improvement error as follows: Under Assumptions 3 and 4, we set the learning rate of PPO $\eta_{PPO} = \min\{(1-\gamma)/3(1+\gamma)^2 1/\sqrt{K_{TD}}\}$, and with a probability of $1 - \delta$:

$$\|(f_{\hat{\theta}} - \tau_{k+1}(\beta^{-1} \tilde{Q}_{\omega_k} + \tau_k^{-1} f_{\theta_k})\|^2$$
$$= \mathcal{O}(\Upsilon^3 m^{-1/2} \log(1/\delta) + \Upsilon^{5/2} m^{-1/4} \sqrt{\log(1/\delta)}$$
$$+ \Upsilon r_{max}^2 m^{-1/4} + \Upsilon^2 K_{TD}^{-1/2} + \Upsilon). \tag{19}$$

We quantify how the errors propagate in neural PPO [31] in the following.

Lemma 7 [31]. *(Error Propagation) We have,*

$$\left| \mathbb{E}_{s \sim \nu_{\pi^*}} \left[\mathbb{E}_{a \sim \pi^*} [\log(\pi_{\theta_{k+1}}/\pi_{k+1})] - \mathbb{E}_{a \sim \pi_{\theta_k}} \right.\right.$$
$$\left.\left. [\log(\pi_{\theta_{k+1}}/\pi_{k+1})] \right| \le \tau_{k+1}^{-1} \varepsilon_k'' \varphi_{k+1}^* + \beta^{-1} \varepsilon_k'' \psi_k^* \right. \tag{20}$$

ε_k'' are defined in Eq. (14) as well as Eq. (19). $\varphi_k^* = \mathbb{E}_{(s,a) \sim \sigma_\pi} \left[\left(\frac{d\pi^*}{d\pi_0} - \right.\right.$
$\left.\left. \frac{d\pi_{\theta_k}}{d\pi_0} \right)^2 \right]^{1/2}, \psi_k^* = \mathbb{E}_{(s,a) \sim \sigma_\pi} \left[\left(\frac{d\sigma_{\pi^*}}{d\sigma_\pi} - \frac{d\nu_{\pi^*}}{d\nu_\pi} \right)^2 \right]^{1/2}. \frac{d\pi^*}{d\pi_0}, \frac{d\pi_{\theta_k}}{d\pi_0}, \frac{d\sigma_{\pi^*}}{d\sigma_\pi}, \frac{d\nu_{\pi^*}}{d\nu_\pi}$ are the Radon-Nikodym derivatives [23]. We denote RHS in Eq. (20) by $\varepsilon_k = \tau_{k+1}^{-1} \varepsilon_k'' \varphi_{k+1}^* + \beta^{-1} \varepsilon_k'' \psi_k^*$. Lemma 7 essentially quantifies the error from which we use the two-layer neural network to approximate the action-value function

and policy instead of having access to the exact ones. Please refer to [31] for complete proofs of Lemma 6 and 7.

$$\left|\mathbb{E}_{s\sim\nu_{\pi^*}}\left[\mathbb{E}_{a\sim\pi^*}\left[\log(\pi_{\theta_{k+1}}/\pi_{k+1})\right.\right.\right.$$
$$\left.\left.\left.-\mathbb{E}_{a\sim\pi_{\theta_k}}\left[\log(\pi_{\theta_{k+1}}/\pi_{k+1})\right]\right]\right|\leq\tau_{k+1}^{-1}\varepsilon_k''\varphi_{k+1}^*+\beta^{-1}\varepsilon_k''\psi_k^*\right.$$

We then characterize the difference between energy functions in each step [31]. Under the optimal policy $\pi*$,

Lemma 8 [31]. *(Stepwise Energy Function difference) Under the same condition of Lemma 7, we have*

$$\mathbb{E}_{s\sim\nu_{\pi^*}}[\|\tau_{k+1}^{-1}f_{\theta_{k+1}}-\tau_k^{-1}f_{\theta_k}\|_\infty^2]\leq 2\varepsilon_k'+2\beta_k^{-2}U, \tag{21}$$

where $\varepsilon_k'=|\mathcal{A}|\tau_{k+1}^{-2}\epsilon_{k+1}^2$
and $U=2\mathbb{E}_{s\sim\nu_{\pi^*}}[\max_{a\in\mathcal{A}}(\tilde{Q}_{\omega_0})^2]+2\Upsilon^2$.

Proof. By the triangle inequality, we get the following,

$$\|\tau_{k+1}^{-1}f_{\theta_{k+1}}-\tau_k^{-1}f_{\theta_k}\|_\infty^2$$
$$\leq 2\left(\|\tau_{k+1}^{-1}f_{\theta_{k+1}}-\tau_k^{-1}f_{\theta_k}-\beta^{-1}\tilde{Q}_{\omega_k}\|_\infty^2+\|\beta^{-1}\tilde{Q}_{\omega_k}\|_\infty^2\right) \tag{22}$$

We take the expectation of both sides of Eq. (22) with respect to $s\sim\nu_{\pi^*}$. With the 1-Lipshitz continuity of $\tilde{Q}_{\omega_k}$ in ω and $\|\omega_k-\Theta_{\text{init}}\|_2\leq\Upsilon$, we have,

$$\mathbb{E}_{\nu_{\pi^*}}[\|\tau_{k+1}^{-1}f_{\theta_{k+1}}-\tau_k^{-1}f_{\theta_k}\|_\infty^2]$$
$$\leq 2(|\mathcal{A}|\tau_{k+1}^{-2}\epsilon_{k+1}^2+\mathbb{E}_{s\sim\nu_{\pi^*}}[\max_{a\in\mathcal{A}}(\tilde{Q}_{\omega_0})^2]+\Upsilon^2)$$

Thus complete the proof.

We then derive a difference term associated with π_{k+1} and π_{θ_k}, where at the k-th iteration π_{k+1} is the solution for the following subproblem,

$$\pi_{k+1}=\arg\max_\pi\left(\mathbb{E}_{s\sim\nu_{\pi_k}}\left[\mathbb{E}_{a\sim\pi}[\tilde{Q}_{\pi_k,\hat{y}_k}]-\beta\text{KL}(\pi\|\pi_{\theta_k})\right]\right)$$

and π_{θ_k} is the policy parameterized by the two-layered over-parameterized neural network. The following lemma establishes the one-step descent of the KL-divergence in the policy space:

Lemma 9 *(One-step difference of π). For π_{k+1} and π_{θ_k}, we have*

$$KL(\pi^*\|\pi_{\theta_k})-KL(\pi^*\|\pi_{\theta_{k+1}})$$
$$\geq\left(\mathbb{E}_{a\sim\pi^*}[\log(\frac{\pi_{\theta_{k+1}}}{\pi_{k+1}})]-\mathbb{E}_{a\sim\pi_{\theta_k}}[\log(\frac{\pi_{\theta_{k+1}}}{\pi_{k+1}})]\right)$$
$$+\beta^{-1}\left(\mathbb{E}_{a\sim\pi^*}[\tilde{Q}_{\pi_k,\hat{y}_k}]-\mathbb{E}_{a\sim\pi_{\theta_k}}[\tilde{Q}_{\pi_k,\hat{y}_k}]\right)$$
$$+\frac{1}{2}\|\pi_{\theta_{k+1}}-\pi_{\theta_k}\|_1^2+\left(\mathbb{E}_{a\sim\pi_{\theta_k}}[\tau_{k+1}^{-1}f_{\theta_{k+1}}-\tau_k^{-1}f_{\theta_k}]\right.$$
$$\left.-\mathbb{E}_{a\sim\pi_{\theta_{k+1}}}[\tau_{k+1}^{-1}f_{\theta_{k+1}}-\tau_k^{-1}f_{\theta_k}]\right) \tag{23}$$

Proof. We start from

$$\mathrm{KL}(\pi^*\|\pi_{\theta_k}) - \mathrm{KL}(\pi^*\|\pi_{\theta_{k+1}}) = \mathbb{E}_{a\sim\pi^*}[\log(\frac{\pi_{\theta_{k+1}}}{\pi_{\theta_k}})]$$

$$(\text{By definition, } \mathrm{KL}(\pi_{\theta_{k+1}}\|\pi_{\theta_k}) = \mathbb{E}_{a\sim\pi_{\theta_{k+1}}}[\log(\frac{\pi_{\theta_{k+1}}}{\pi_{\theta_k}})]))$$

$$= \left(\mathbb{E}_{a\sim\pi^*}[\log(\frac{\pi_{\theta_{k+1}}}{\pi_{\theta_k}})] - \mathbb{E}_{a\sim\pi_{\theta_{k+1}}}[\log(\frac{\pi_{\theta_{k+1}}}{\pi_{\theta_k}})]\right)+$$

$$\mathrm{KL}(\pi_{\theta_{k+1}}\|\pi_{\theta_k})$$

We then add and subtract terms,

$$= \mathbb{E}_{a\sim\pi^*}[\log(\frac{\pi_{\theta_{k+1}}}{\pi_{\theta_k}})] - \mathbb{E}_{a\sim\pi_{\theta_{k+1}}}[\log(\frac{\pi_{\theta_{k+1}}}{\pi_{\theta_k}})] + \mathrm{KL}$$

$$(\pi_{\theta_{k+1}}\|\pi_{\theta_k}) + \beta^{-1}\left(\mathbb{E}_{a\sim\pi^*}[\tilde{Q}_{\pi_k,\hat{y}_k}] - \mathbb{E}_{a\sim\pi_{\theta_k}}[\tilde{Q}_{\pi_k,\hat{y}_k}]\right)$$

$$- \beta^{-1}\left(\mathbb{E}_{a\sim\pi^*}[\tilde{Q}_{\pi_k,\hat{y}_k}] - \mathbb{E}_{a\sim\pi_{\theta_k}}[\tilde{Q}_{\pi_k,\hat{y}_k}]\right)$$

$$+ \mathbb{E}_{a\sim\pi_{\theta_k}}[\log(\frac{\pi_{\theta_{k+1}}}{\pi_{\theta_k}})] - \mathbb{E}_{a\sim\pi_{\theta_k}}[\log(\pi_{\theta_{k+1}}\pi_{\theta_k})]$$

Rearrange the terms and we get,

$$= \left(\mathbb{E}_{a\sim\pi^*}[\log(\pi_{\theta_{k+1}}) - \log(\pi_{\theta_k}) - \beta^{-1}\tilde{Q}_{\pi_k,\hat{y}_k}]\right.$$

$$- \mathbb{E}_{a\sim\pi_{\theta_k}}[\log(\pi_{\theta_{k+1}}) - \log(\pi_{\theta_k}) - \beta^{-1}\tilde{Q}_{\pi_k,\hat{y}_k}]\right)$$

$$+ \beta^{-1}\left(\mathbb{E}_{a\sim\pi^*}[\tilde{Q}_{\pi_k,\hat{y}_k}] - \mathbb{E}_{a\sim\pi_{\theta_k}}[\tilde{Q}_{\pi_k,\hat{y}_k}]\right) + \mathrm{KL}$$

$$(\pi_{\theta_{k+1}}\|\pi_{\theta_k}) + \left(\mathbb{E}_{a\sim\pi_{\theta_k}}[\log(\frac{\pi_{\theta_{k+1}}}{\pi_{\theta_k}})] - \mathbb{E}_{a\sim\pi_{\theta_{k+1}}}\right.$$

$$[\log(\pi_{\theta_{k+1}}\pi_{\theta_k})]\right) \tag{24}$$

Recall that $\pi_{k+1} \propto \exp\{\tau_k^{-1}f_{\theta_k} + \beta^{-1}\tilde{Q}_{\pi_k}^y\}$. We define the two normalization factors associated with ideal improved policy π_{k+1} and the current parameterized policy π_{θ_k} as,

$$Z_{k+1}(s) := \sum_{a'\in\mathcal{A}} \exp\{\tau_k^{-1}f_{\theta_k}(s,a') + \beta^{-1}\tilde{Q}_{\pi_k}^y(s,a')\}$$

$$Z_{\theta_{k+1}}(s) := \sum_{a'\in\mathcal{A}} \exp\{\tau_{k+1}^{-1}f_{\theta_{k+1}}(s,a')\}$$

We then have,

$$\pi_{k+1}(a|s) = \frac{\exp\{\tau_k^{-1}f_{\theta_k}(s,a) + \beta^{-1}\tilde{Q}_{\pi_k}^y(s,a)\}}{Z_{k+1}(s)}, \tag{25}$$

$$\pi_{\theta_{k+1}}(a|s) = \frac{\exp\{\tau_{k+1}^{-1}f_{\theta_{k+1}}(s,a)\}}{Z_{\theta_{k+1}}(s)} \tag{26}$$

For any π, π' and k, we have,

$$\mathbb{E}_{a\sim\pi}[\log Z_{\theta_{k+1}}] - \mathbb{E}_{a\sim\pi'}[\log Z_{\theta_{k+1}}] = 0 \tag{27}$$

$$\mathbb{E}_{a\sim\pi}[\log Z_{k+1}] - \mathbb{E}_{a\sim\pi'}[\log Z_{k+1}] = 0 \qquad (28)$$

Now we look back at a few terms on RHS from Eq. (24):

$$
\begin{aligned}
&\mathbb{E}_{a\sim\pi^*}\big[\log(\pi_{\theta_k}) + \beta^{-1}\tilde{Q}_{\pi_k,\hat{y}_k}\big] \\
&\quad - \mathbb{E}_{a\sim\pi_{\theta_k}}\big[\log(\pi_{\theta_k}) + \beta^{-1}\tilde{Q}_{\pi_k,\hat{y}_k}\big] \\
&= \big(\mathbb{E}_{a\sim\pi^*}[\tau_k^{-1}f_{\theta_k} + \beta^{-1}\tilde{Q}_{\pi_k,\hat{y}_k} - \log Z_{\theta_{k+1}}] \\
&\quad - \mathbb{E}_{a\sim\pi_{\theta_k}}[\tau_k^{-1}f_{\theta_k} + \beta^{-1}\tilde{Q}_{\pi_k,\hat{y}_k} - \log Z_{\theta_{k+1}}]\big) \\
&= \mathbb{E}_{a\sim\pi^*}\Big[\log\frac{\exp\{\tau_k^{-1}f_{\theta_k} + \beta^{-1}\tilde{Q}_{\pi_k,\hat{y}_k}\}}{Z_{k+1}}\Big] \\
&\quad - \mathbb{E}_{a\sim\pi_{\theta_k}}\Big[\log\frac{\exp\{\tau_k^{-1}f_{\theta_k} + \beta^{-1}\tilde{Q}_{\pi_k,\hat{y}_k}\}}{Z_{k+1}}\Big] \\
&= \mathbb{E}_{a\sim\pi^*}[\log\pi_{k+1}] - \mathbb{E}_{a\sim\pi_{\theta_k}}[\log\pi_{k+1}] \qquad (29)
\end{aligned}
$$

For Eq. (29), we obtain the first equality by Eq. (26). Then, by swapping Eq. (27) with Eq. (28), we obtain the second equality. We achieve the concluding step with the definition in Eq. (25). Following a similar logic, we have,

$$
\begin{aligned}
&\mathbb{E}_{a\sim\pi_{\theta_k}}[\log(\frac{\pi_{\theta_{k+1}}}{\pi_{\theta_k}})] - \mathbb{E}_{a\sim\pi_{\theta_{k+1}}}[\log(\frac{\pi_{\theta_{k+1}}}{\pi_{\theta_k}})] \\
&= \mathbb{E}_{a\sim\pi_{\theta_k}}[\tau_{k+1}^{-1}f_{\theta_{k+1}} - \log Z_{\theta_{k+1}} - \tau_k^{-1}f_{\theta_k} + \log Z_{\theta_k}] - \\
&\quad \mathbb{E}_{a\sim\pi_{\theta_{k+1}}}[\tau_{k+1}^{-1}f_{\theta_{k+1}} - \log Z_{\theta_{k+1}} - \tau_k^{-1}f_{\theta_k} + \log Z_{\theta_k}] \\
&= \mathbb{E}_{a\sim\pi_{\theta_k}}[\tau_{k+1}^{-1}f_{\theta_{k+1}} - \tau_k^{-1}f_{\theta_k}] - \mathbb{E}_{a\sim\pi_{\theta_{k+1}}}[\tau_{k+1}^{-1}f_{\theta_{k+1}} - \\
&\quad \tau_k^{-1}f_{\theta_k}] \qquad (30)
\end{aligned}
$$

Finally, by using the Pinsker's inequality [12], we have,

$$\mathrm{KL}(\pi_{\theta_{k+1}}\|\pi_{\theta_k}) \geq 1/2\|\pi_{\theta_{k+1}} - \pi_{\theta_k}\|_1^2 \qquad (31)$$

Plugging Eqs. (29), (30), and (31) into Eq. (24), we have

$$
\begin{aligned}
&\mathrm{KL}(\pi^*\|\pi_{\theta_k}) - \mathrm{KL}(\pi^*\|\pi_{\theta_{k+1}}) \\
&\geq \big(\mathbb{E}_{a\sim\pi^*}[\log(\pi_{\theta_{k+1}}) - \log(\pi_{k+1})] - \mathbb{E}_{a\sim\pi_{\theta_k}}[\log(\pi_{\theta_{k+1}}) \\
&\quad - \log(\pi_{k+1})]\big) + \beta^{-1}\big(\mathbb{E}_{a\sim\pi^*}[\tilde{Q}_{\pi_k,\hat{y}_k}] - \mathbb{E}_{a\sim\pi_{\theta_k}}[\tilde{Q}_{\pi_k,\hat{y}_k}]\big) \\
&\quad + \frac{1}{2}\|\pi_{\theta_{k+1}} - \pi_{\theta_k}\|_1^2 + \big(\mathbb{E}_{a\sim\pi_{\theta_k}}[\tau_{k+1}^{-1}f_{\theta_{k+1}} - \tau_k^{-1}f_{\theta_k}] \\
&\quad - \mathbb{E}_{a\sim\pi_{\theta_{k+1}}}[\tau_{k+1}^{-1}f_{\theta_{k+1}} - \tau_k^{-1}f_{\theta_k}]\big)
\end{aligned}
$$

Rearranging the terms, we obtain Lemma 9.

Lemma 9 serves as an intermediate-term for the major result's proof. We obtain upper bounds by telescoping this term in Theorem 1. Now we are ready to present the proof for Theorem 1.

Proof. First we take expectation of both sides of Eq. (23) with respect to $s \sim \nu_{\pi^*}$ from Lemma 9 and insert Eq. (20) to obtain,

$$
\begin{aligned}
&\mathbb{E}_{s\sim\nu_{\pi^*}}[\mathrm{KL}(\pi^*\|\pi_{\theta_{k+1}})] - \mathbb{E}_{s\sim\nu_{\pi^*}}[\mathrm{KL}(\pi^*\|\pi_{\theta_k})] \\
&\leq \varepsilon_k - \beta^{-1}\mathbb{E}_{s\sim\nu_{\pi^*}}\big[\mathbb{E}_{a\sim\pi^*}[\tilde{Q}_{\pi_k,\hat{y}_k}] - \mathbb{E}_{a\sim\pi_{\theta_k}}[\tilde{Q}_{\pi_k,\hat{y}_k}]\big] \\
&\quad - 1/2\mathbb{E}_{s\sim\nu_{\pi^*}}\big[\|\pi_{\theta_{k+1}} - \pi_{\theta_k}\|_1^2\big] - \mathbb{E}_{s\sim\nu_{\pi^*}}\big[\mathbb{E}_{a\sim\pi_{\theta_k}} \\
&\quad [\tau_{k+1}^{-1}f_{\theta_{k+1}} - \tau_k^{-1}f_{\theta_k}] - \mathbb{E}_{a\sim\pi_{\theta_{k+1}}}[\tau_{k+1}^{-1}f_{\theta_{k+1}} - \tau_k^{-1}f_{\theta_k}]\big]
\end{aligned}
\tag{32}
$$

Then, by Lemma 3, we have,

$$
\begin{aligned}
&\beta^{-1}\mathbb{E}_{s\sim\nu_{\pi^*}}\big[\mathbb{E}_{a\sim\pi^*}[\tilde{Q}_{\pi_k,\hat{y}_k}] - \mathbb{E}_{a\sim\pi_{\theta_k}}[\tilde{Q}_{\pi_k,\hat{y}_k}]\big] \\
&= \beta^{-1}(1-\gamma)\big(J_\lambda^{\hat{y}_k}(\pi^*) - J_\lambda^{\hat{y}_k}(\pi)\big)
\end{aligned}
\tag{33}
$$

And with Hölder's inequality, we have,

$$
\begin{aligned}
&\mathbb{E}_{s\sim\nu_{\pi^*}}\big[\mathbb{E}_{a\sim\pi_{\theta_k}}[\tau_{k+1}^{-1}f_{\theta_{k+1}} - \tau_k^{-1}f_{\theta_k}] - \mathbb{E}_{a\sim\pi_{\theta_{k+1}}} \\
&[\tau_{k+1}^{-1}f_{\theta_{k+1}} - \tau_k^{-1}f_{\theta_k}]\big] \\
&= \mathbb{E}_{s\sim\nu_{\pi^*}}\big[\langle\tau_{k+1}^{-1}f_{\theta_{k+1}} - \tau_k^{-1}f_{\theta_k}, \pi_{\theta_k} - \pi_{\theta_{k+1}}\rangle\big] \\
&\leq \mathbb{E}_{s\sim\nu_{\pi^*}}\big[\|\tau_{k+1}^{-1}f_{\theta_{k+1}} - \tau_k^{-1}f_{\theta_k}\|_\infty\|\pi_{\theta_k} - \pi_{\theta_{k+1}}\|_1\big]
\end{aligned}
\tag{34}
$$

Insert Eqs. (33) and (34) into Eq. (32), we have,

$$
\begin{aligned}
&\mathbb{E}_{s\sim\nu_{\pi^*}}[\mathrm{KL}(\pi^*\|\pi_{\theta_{k+1}})] - \mathbb{E}_{s\sim\nu_{\pi^*}}[\mathrm{KL}(\pi^*\|\pi_{\theta_k})] \\
&\leq \varepsilon_k - (1-\gamma)\beta^{-1}\big(J_\lambda^{\hat{y}_k}(\pi^*) - J_\lambda^{\hat{y}_k}(\pi)\big) - 1/2\mathbb{E}_{s\sim\nu_{\pi^*}} \\
&\quad \big[\|\pi_{\theta_{k+1}} - \pi_{\theta_k}\|_1^2\big] + \mathbb{E}_{s\sim\nu_{\pi^*}}\big[\|\tau_{k+1}^{-1}f_{\theta_{k+1}} - \tau_k^{-1}f_{\theta_k}\|_\infty \\
&\quad \|\pi_{\theta_k} - \pi_{\theta_{k+1}}\|_1\big] \\
&\leq \varepsilon_k - (1-\gamma)\beta^{-1}\big(J_\lambda^{y^*}(\pi^*) - J_\lambda^{\hat{y}_k}(\pi) - J_\lambda^{y^*}(\pi^*) \\
&\quad + J_\lambda^{\hat{y}_k}(\pi^*)\big) + 1/2\mathbb{E}_{s\sim\nu_{\pi^*}}\big[\|\tau_{k+1}^{-1}f_{\theta_{k+1}} - \tau_k^{-1}f_{\theta_k}\|_\infty^2\big] \\
&\leq \varepsilon_k - (1-\gamma)\beta^{-1}\big(J_\lambda^{y^*}(\pi^*) - J_\lambda^{\hat{y}_k}(\pi)\big) \\
&\quad + (1-\gamma)\beta^{-1}\big(J_\lambda^{y^*}(\pi^*) - J_\lambda^{\hat{y}_k}(\pi^*)\big) \\
&\quad + 1/2\mathbb{E}_{s\sim\nu_{\pi^*}}\big[\|\tau_{k+1}^{-1}f_{\theta_{k+1}} - \tau_k^{-1}f_{\theta_k}\|_\infty^2\big].
\end{aligned}
$$

The second inequality holds by using the inequality $2AB - B^2 \leq A^2$, with a minor abuse of notations. Here, $A := \|\tau_{k+1}^{-1}f_{\theta_{k+1}} - \tau_k^{-1}f_{\theta_k}\|_\infty$ and $B := \|\pi_{\theta_k} - \pi_{\theta_{k+1}}\|_1$. Then, by plugging in Lemma 4 and Eq. (21) we end up with,

$$
\begin{aligned}
&\mathbb{E}_{s\sim\nu_{\pi^*}}[\mathrm{KL}(\pi^*\|\pi_{\theta_{k+1}})] - \mathbb{E}_{s\sim\nu_{\pi^*}}[\mathrm{KL}(\pi^*\|\pi_{\theta_k})] \\
&\leq \varepsilon_k - (1-\gamma)\beta^{-1}\big(J_\lambda^{y^*}(\pi^*) - J_\lambda^{\hat{y}_k}(\pi_k)\big) \\
&\quad + (1-\gamma)\beta^{-1}\Big(\frac{2c_3 M(1-\gamma)\lambda}{\sqrt{k}}\Big) + (\varepsilon_k' + \beta_k^{-2}U)
\end{aligned}
\tag{35}
$$

Rearrange Eq. (35), we have

$$(1-\gamma)\beta^{-1}\big(J_\lambda^{y^*}(\pi^*) - J_\lambda^{\hat{y}_k}(\pi_k)\big)$$
$$\leq \mathbb{E}_{s\sim\nu_{\pi^*}}[\mathrm{KL}(\pi^*\|\pi_{\theta_k})] - \mathbb{E}_{s\sim\nu_{\pi^*}}[\mathrm{KL}(\pi^*\|\pi_{\theta_{k+1}})]$$
$$+ \big(\frac{2c_3 M(1-\gamma)^2\lambda}{\beta\sqrt{k}}\big) + \varepsilon_k + \varepsilon_k' + \beta_k^{-2}U \tag{36}$$

And then telescoping Eq. (36) results in,

$$(1-\gamma)\sum_{k=1}^{K}\beta^{-1}\min_{k\in[K]}\big(J_\lambda^{y^*}(\pi^*) - J_\lambda^{\hat{y}_k}(\pi_k)\big)$$

$$\leq (1-\gamma)\sum_{k=1}^{K}\beta^{-1}\big(J_\lambda^{y^*}(\pi^*) - J_\lambda^{\hat{y}_k}(\pi_k)\big)$$

$$\leq \mathbb{E}_{s\sim\nu_{\pi^*}}[\mathrm{KL}(\pi^*\|\pi_0)] - \mathbb{E}_{s\sim\nu_{\pi^*}}[\mathrm{KL}(\pi^*\|\pi_K)]$$

$$+ \lambda r_{\max}(1-\gamma)^2\sum_{k=1}^{K}\beta^{-1}\big(\frac{2c_3}{\sqrt{k}}\big) + U\sum_{k=1}^{K}\beta_k^{-2}$$

$$+ \sum_{k=1}^{K}(\varepsilon_k + \varepsilon_k') \tag{37}$$

We complete the final step in Eq. (37) by plugging in Lemma 4 and Eq. (20). Per the observation we make in the proof of Theorem 2,

1. $\mathbb{E}_{s\sim\nu_{\pi^*}}[\mathrm{KL}(\pi^*\|\pi_0)] \leq \log\mathcal{A}$ due to the uniform initialization of policy.
2. $\mathrm{KL}(\pi^*\|\pi_K)$ is a non-negative term.

We now have,

$$\min_{k\in[K]} J_\lambda^{y^*}(\pi^*) - J_\lambda^{\hat{y}_k}(\pi_k)$$

$$\leq \frac{\log|\mathcal{A}| + UK\beta^{-2} + \sum_{k=1}^{K}(\varepsilon_k + \varepsilon_k')}{(1-\gamma)K\beta^{-1}})$$

$$+ \lambda r_{\max}(1-\gamma)\big(\frac{2c_3}{\sqrt{k}}\big)$$

Replacing β with $\beta_0\sqrt{K}$ finishes the proof.

D.4   Proof of Theorem 2

In the following part, we focus the convergence of neural NPG. We first define the following terms under neural NPG update rule.

Lemma 10 [51]. *For energy-based policy π_θ, we have policy gradient and Fisher information matrix,*

$$\nabla_\theta J(\pi_\theta) = \tau \mathbb{E}_{d_{\pi_\theta}(s,a)}[Q_{\pi_\theta}(s,a)(\phi_\theta(s,a) - \mathbb{E}_{\pi_\theta}[\phi_\theta(s,a')])]$$

$$F(\theta) = \tau^2 \mathbb{E}_{d_{\pi_\theta}(s,a)}[(\phi_\theta(s,a) - \mathbb{E}_{\pi_\theta}[\phi_\theta(s,a')])$$
$$(\phi_\theta(s,a) - \mathbb{E}_{\pi_\theta}[\phi_\theta(s,a')])^\top]$$

We then derive an upper bound for $J_\lambda^{y^*}(\pi^*) - J_\lambda^{y^*}(\pi_k)$ for the neural NPG method in the following lemma:

Lemma 11 *(One-step difference of π). It holds that, with probability of $1 - \delta$,*

$$(1-\gamma)\left(J_\lambda^{\hat{y}_k}(\pi^*) - J_\lambda^{\hat{y}_k}(\pi_k)\right)$$
$$\leq \eta_{\mathrm{NPG}}^{-1} \mathbb{E}_{s \sim \nu_{\pi^*}}\left[KL(\pi^*\|\pi_k) - KL(\pi^*\|\pi_{k+1})\right]$$
$$+ \eta_{\mathrm{NPG}}(9\Upsilon^2 + r_{\max}^2) + 2c_0\epsilon_k' + \eta_{\mathrm{NPG}}^{-1}\epsilon_k'',$$

where

$$\epsilon_k' = \mathcal{O}(\Upsilon^3 m^{-1/2}\log(1/\delta) + \Upsilon^{5/2}m^{-1/4}\sqrt{\log(1/\delta)}$$
$$+ \Upsilon r_{\max}^2 m^{-1/4} + \Upsilon^2 K_{\mathrm{TD}}^{-1/2} + \Upsilon),$$
$$\epsilon_k'' = 8\eta_{\mathrm{NPG}}\Upsilon^{1/2}c_0\sigma_\xi^{1/2}T^{-1/4}$$
$$+ \mathcal{O}((\tau_{k+1} + \eta_{\mathrm{NPG}})\Upsilon^{3/2}m^{-1/4}$$
$$+ \eta_{\mathrm{NPG}}\Upsilon^{5/4}m^{-1/8}),$$

c_0 is defined in Assumption 2 and σ_ξ is defined in Assumption 1. Meanwhile, Υ is the radius of the parameter space, m is the width of the neural network, and T is the sample batch size.

Proof. We start from the following,

$$KL(\pi^*\|\pi_k) - KL(\pi^*\|\pi_{k+1}) - KL(\pi_{k+1}\|\pi_k)$$
$$= \mathbb{E}_{a \sim \pi^*}\left[\log(\frac{\pi_{k+1}}{\pi_k})\right] - \mathbb{E}_{a \sim \pi_{k+1}}\left[\log(\frac{\pi_{k+1}}{\pi_k})\right] \qquad (38)$$

(by KL's definition).

We now show the building blocks of the proof. *First*, we add and subtract a few terms to RHS of Eq. (38) then take the expectation of both sides with respect to $s \sim \nu_{\pi^*}$. Rearrange these terms, we get,

$$\mathbb{E}_{s \sim \nu_{\pi^*}}\left[KL(\pi^*\|\pi_k) - KL(\pi^*\|\pi_{k+1}) - KL(\pi_{k+1}\|\pi_k)\right]$$
$$= \eta_{\mathrm{NPG}}\mathbb{E}_{s \sim \nu_{\pi^*}}\left[\mathbb{E}_{a \sim \pi^*}[\tilde{Q}_{\pi_k,\hat{y}_k}] - \mathbb{E}_{a \sim \pi_k}[\tilde{Q}_{\pi_k,\hat{y}_k}]\right]$$
$$+ H_k \qquad (39)$$

where H_k is denoted by,

$$H_k := \mathbb{E}_{s \sim \nu_{\pi^*}} \big[\mathbb{E}_{a \sim \pi^*} [\log(\frac{\pi_{k+1}}{\pi_k}) - \eta_{\mathrm{NPG}} \tilde{Q}_{\omega_k}]$$
$$- \mathbb{E}_{a \sim \pi_k} \big[\log(\frac{\pi_{k+1}}{\pi_k}) - \eta_{\mathrm{NPG}} \tilde{Q}_{\omega_k} \big]$$
$$+ \eta_{\mathrm{NPG}} \mathbb{E}_{s \sim \nu_{\pi^*}} \big[\mathbb{E}_{a \sim \pi^*} [\tilde{Q}_{\omega_k} - \tilde{Q}_{\pi_k, \hat{y}_k}]$$
$$- \mathbb{E}_{a \sim \pi_k} [\tilde{Q}_{\omega_k} - \tilde{Q}_{\pi_k, \hat{y}_k}]]$$
$$+ \mathbb{E}_{s \sim \nu_{\pi^*}} \big[\mathbb{E}_{a \sim \pi_k} [\log(\frac{\pi_{k+1}}{\pi_k})]$$
$$- \mathbb{E}_{a \sim \pi_{k+1}} [\log(\frac{\pi_{k+1}}{\pi_k})]] \big] \tag{40}$$

By Lemma 3, we have

$$\eta_{\mathrm{NPG}} \mathbb{E}_{s \sim \nu_{\pi^*}} \big[\mathbb{E}_{a \sim \pi^*} [\tilde{Q}_{\pi_k, \hat{y}_k}] - \mathbb{E}_{a \sim \pi_k} [\tilde{Q}_{\pi_k, \hat{y}_k}] \big]$$
$$= \eta_{\mathrm{NPG}} (1 - \gamma) \big(J_\lambda^{\hat{y}_k}(\pi^*) - J_\lambda^{\hat{y}_k}(\pi_k) \big) \tag{41}$$

Insert Eqs. (41) back to Eq. (39), we have,

$$\eta_{\mathrm{NPG}} (1 - \gamma) \big(J_\lambda^{\hat{y}_k}(\pi^*) - J_\lambda^{\hat{y}_k}(\pi_k) \big)$$
$$= \mathbb{E}_{s \sim \nu_{\pi^*}} \big[\mathrm{KL}(\pi^* \| \pi_k) - \mathrm{KL}(\pi^* \| \pi_{k+1}) - \mathrm{KL}(\pi_{k+1} \| \pi_k) \big]$$
$$- H_k$$
$$\leq \mathbb{E}_{s \sim \nu_{\pi^*}} \big[\mathrm{KL}(\pi^* \| \pi_k) - \mathrm{KL}(\pi^* \| \pi_{k+1}) - \mathrm{KL}(\pi_{k+1} \| \pi_k) \big]$$
$$+ |H_k| \tag{42}$$

We reach the final inequality of Eq. (42) by algebraic manipulation. *Second*, we follow Lemma 5.5 of [51] and obtain an upper bound for Eq. (40). Specifically, with probability of $1 - \delta$,

$$\mathbb{E}_{a \sim \mathrm{init}} \Big[|H_k| - \mathbb{E}_{s \sim \nu_{\pi^*}} [\mathrm{KL}(\pi_{k+1} \| \pi_k)] \Big]$$
$$\leq \eta_{\mathrm{NPG}}^2 (9\Upsilon^2 + r_{\max}^2) + 2\eta_{\mathrm{NPG}} c_0 \epsilon_k' + \epsilon_k'' \tag{43}$$

The expectation is taken over randomness. With these building blocks of Eqs. (42) and (43), we are now ready to reach the concluding inequality. Plugging Eqs. (43) back into Eq. (42), we end up with, with probability of $1 - \delta$,

$$\eta_{\mathrm{NPG}} (1 - \gamma) \big(J_\lambda^{\hat{y}_k}(\pi^*) - J_\lambda^{\hat{y}_k}(\pi_k) \big)$$
$$\leq \mathbb{E}_{s \sim \nu_{\pi^*}} \big[\mathrm{KL}(\pi^* \| \pi_k) - \mathrm{KL}(\pi^* \| \pi_{k+1}) \big]$$
$$+ \eta_{\mathrm{NPG}}^2 (9\Upsilon^2 + r_{\max}^2) + 2\eta_{\mathrm{NPG}} c_0 \epsilon_k' + \epsilon_k'' \tag{44}$$

Dividing both sides of Eq. (44) by η_{NPG} completes the proof. The details are included in the Appendix.

We have the following Lemma to bound the error terms H_k defined in Eq. (40) of Lemma 11.

Lemma 12 [51]. *Under Assumptions 4, we have*

$$\mathbb{E}_{a\sim\text{init}}\Big[|H_k| - \mathbb{E}_{s\sim\nu_{\pi^*}}[KL(\pi_{k+1}\|\pi_k)]\Big]$$
$$\leq \eta_{\text{NPG}}^2(9\Upsilon^2 + r_{\max}^2) + \eta_{\text{NPG}}(\varphi_k' + \psi_k')\epsilon_k' + \epsilon_k''$$

Here the expectation is taken over all the randomness. We have $\epsilon_k' := \|Q_{\omega_k} - Q_{\pi_k}\|_{\nu_{\pi_k}}^2$ *and*

$$\epsilon_k'' = \sqrt{2}\Upsilon^{1/2}\eta_{\text{NPG}}(\varphi_k + \psi_k)\tau_k^{-1}\big\{\mathbb{E}_{(s,a)\sim\sigma_{\pi\theta_k}}[\|\xi_k(\delta_k)\|_2^2]$$
$$+ \mathbb{E}_{(s,a)\sim\sigma_{\pi\omega_k}}[\|\xi_k(\omega_k)\|_2^2]\big\}^{1/2}$$
$$+ \mathcal{O}((\tau_{k+1} + \eta_{\text{NPG}})\Upsilon^{3/2}m^{-1/4} + \eta_{\text{NPG}}\Upsilon^{5/4}m^{-1/8}).$$

Recall $\xi_k(\omega_k)$ *and* $\xi_k(\omega_k)$ *are defined in Assumption 1, while* φ_k, ψ_k, φ_k', *and* ψ_k *are defined in Assumption 2.*

Please refer to [51] for complete proof. Finally, we are ready to show the proof for Theorem 2.

Proof. First, we combine Lemma 4 and 11 to get the following:

$$(1-\gamma)\big(J_\lambda^{y^*}(\pi^*) - J_\lambda^{\hat{y}_k}(\pi^*) + J_\lambda^{\hat{y}_k}(\pi^*) - J_\lambda^{\hat{y}_k}(\pi_k)\big)$$
$$\leq \eta_{\text{NPG}}^{-1}\mathbb{E}_{s\sim\nu_{\pi^*}}[KL(\pi^*\|\pi_k) - KL(\pi^*\|\pi_{k+1})]$$
$$+ \eta_{\text{NPG}}(9\Upsilon^2 + r_{\max}^2) + 2c_0\epsilon_k' + \eta_{\text{NPG}}^{-1}\epsilon_k''$$
$$+ \frac{2c_3 M(1-\gamma)^2\lambda}{\sqrt{k}} \tag{45}$$

We can then see this:

1. $\mathbb{E}_{s\sim\nu_{\pi^*}}[KL(\pi^*\|\pi_1)] \leq \log|\mathcal{A}|$ due to the uniform initialization of policy.
2. $KL(\pi^*\|\pi_{K+1})$ is a non-negative term.

And by setting $\eta_{\text{NPG}} = 1/\sqrt{K}$ and telescoping Eq. (45), we obtain,

$$(1-\gamma)\min_{k\in[K]}\big(J_\lambda^{y^*}(\pi^*) - J_\lambda^{\hat{y}_k}(\pi_k)\big)$$
$$\leq (1-\gamma)\frac{1}{K}\sum_{k=1}^{K}\mathbb{E}(J_\lambda^{y^*}(\pi^*) - J_\lambda^{\hat{y}_k}(\pi_k))$$
$$\leq \frac{1}{\sqrt{K}}(\mathbb{E}_{s\sim\nu_{\pi^*}}[KL(\pi^*\|\pi_1)] + 9\Upsilon^2 + r_{\max}^2) + \frac{1}{K}\sum_{k=1}^{K}$$
$$(2\sqrt{K}c_0\epsilon_k' + \eta_{\text{NPG}}^{-1}\epsilon_k'' + \frac{2c_3 M(1-\gamma)^2\lambda}{\sqrt{k}}) \tag{46}$$

plug ϵ_k' and ϵ_k'' defined in Lemma 11 into Eq. (46), and set ϵ_k as,

$$
\begin{aligned}
\epsilon_k = {}& \sqrt{8}c_0 \Upsilon^{1/2}\sigma_\xi^{1/2}T^{-1/4} \\
& + \mathcal{O}\big((\tau_{k+1}K^{1/2}+1)\Upsilon^{3/2}m^{-1/4} + \Upsilon^{5/4}m^{-1/8}\big) \\
& + c_0\mathcal{O}\big(\Upsilon^3 m^{-1/2}\log(1/\delta) + \Upsilon^{5/2}m^{-1/4}\sqrt{\log(1/\delta)} \\
& + \Upsilon r_{\max}^2 m^{-1/4} + \Upsilon^2 K_{\mathrm{TD}}^{-1/2} + \Upsilon\big)
\end{aligned}
$$

we complete the proof.

D.5   Proof of Lemma 1

Proof. First, we have $\mathbb{E}[G] = \frac{1}{1-\gamma}\mathbb{E}[R]$, i.e., the per-step reward R is an unbiased estimator of the cumulative reward G. Second, it is proved that $\mathbb{V}(G) \le \frac{\mathbb{V}(R)}{(1-\gamma)^2}$ [7]. Given $\lambda \ge 0$, summing up the above equality and inequality, we have

$$
\begin{aligned}
\frac{1}{(1-\gamma)}J_{\frac{\lambda}{(1-\gamma)}}(\pi) &= \frac{1}{(1-\gamma)}\left(\mathbb{E}[R] - \frac{\lambda}{(1-\gamma)}\mathbb{V}(R)\right) \\
&\le \mathbb{E}[G] - \lambda\mathbb{V}(G) = J_\lambda^G(\pi).
\end{aligned}
$$

It completes the proof.

D.6   Proof of Lemma 2

We first provide the supporting lemmas for Lemma 2. We define the local linearization of $f((s,a);\theta)$ defined in Eq. (4) at the initial point Θ_{init} as,

$$
\hat{f}((s,a);\theta) = \frac{1}{\sqrt{m}}\sum_{v=1}^m b_v \mathbb{1}\{[\Theta_{\mathrm{init}}]_v^\top(s,a) > 0\}[\theta]_v^\top(s,a) \tag{47}
$$

We then define the following function spaces,

$$
\mathcal{F}_{\Upsilon,m} := \left\{ \frac{1}{\sqrt{m}}\sum_{v=1}^m b_v \mathbb{1}\{[\Theta_{\mathrm{init}}]_v^\top(s,a) > 0\}[\theta]_v^\top(s,a) : \right.
$$
$$
\left. \|\theta - \Theta_{\mathrm{init}}\|_2 \le \Upsilon \right\},
$$

and

$$
\bar{\mathcal{F}}_{\Upsilon,m} := \left\{ \frac{1}{\sqrt{m}}\sum_{v=1}^m b_v \mathbb{1}\{[\Theta_{\mathrm{init}}]_v^\top(s,a) > 0\}[\theta]_v^\top(s,a) : \right.
$$
$$
\left. \|[\theta]_v - [\Theta_{\mathrm{init}}]_v\|_\infty \le \Upsilon/\sqrt{md} \right\}.
$$

$[\Theta_{\mathrm{init}}]_r \sim \mathcal{N}(0, I_d/d)$ and $b_r \sim \mathrm{Unif}(\{-1,1\})$ are the initial parameters. By the definition, $\bar{\mathcal{F}}_{\Upsilon,m}$ is a subset of $\mathcal{F}_{\Upsilon,m}$. The following lemma characterizes the deviation of $\bar{\mathcal{F}}_{\Upsilon,m}$ from $\mathcal{F}_{\Upsilon,\infty}$.

Lemma 13 *(Projection Error)* [40]. *Let $f \in \mathcal{F}_{\Upsilon,\infty}$, where $\mathcal{F}_{\Upsilon,\infty}$ is defined in Assumption 3. For any $\delta > 0$, it holds with probability at least $1 - \delta$ that*

$$\|\Pi_{\tilde{\mathcal{F}}_{\Upsilon,m}} f - f\|_{\varsigma} \leq \Upsilon m^{-1/2}[1 + \sqrt{2\log(1/\delta)}]$$

where ς is any distribution over $S \times A$.

Please refer to [40] for a detail proof.

Lemma 14 *(Linearization Error). Under Assumption 4, for all $\theta \in \mathcal{D}$, where $\mathcal{D} = \{\xi \in \mathbb{R}^{md} : \|\xi - \Theta_{init}\|_2 \leq \Upsilon\}$, it holds that,*

$$\mathbb{E}_{\nu_\pi}\left[\left(f((s,a);\theta) - \hat{f}((s,a);\theta)\right)^2\right] \leq \frac{4c_1\Upsilon^3}{\sqrt{m}}$$

where $c_1 = c\sqrt{\mathbb{E}_{\mathcal{N}(0,I_d/d)}[1/\|(s,a)\|_2^2]}$, and c is defined in Assumption 4.

Proof. We start from the definitions in Eq. (4) and Eq. (47),

$$\mathbb{E}_{\nu_\pi}\left[\left(f((s,a);\theta) - \hat{f}((s,a);\theta)\right)^2\right]$$

$$= \mathbb{E}_{\nu_\pi}\left[\left(\frac{1}{\sqrt{m}}\Big|\sum_{v=1}^m \left((\mathbb{1}\{[\theta]_v^\top(s,a) > 0\} - \mathbb{1}\{[\Theta_{init}]_v^\top(s,a)\right.\right.\right.$$

$$\left.\left.\left. > 0\})b_v[\theta]_v^\top(s,a)\right)\Big|\right)^2\right]$$

$$\leq \frac{1}{m}\mathbb{E}_{\nu_\pi}\left[\left(\sum_{v=1}^m \left(\Big|\mathbb{1}\{[\theta]_v^\top(s,a) > 0\} - \mathbb{1}\{[\Theta_{init}]_v^\top(s,a)\right.\right.\right.$$

$$\left.\left.\left. > 0\}\Big|\|b_v\|\big|[\theta]_v^\top(s,a)\big|\right)\right)^2\right] \tag{48}$$

The above inequality holds because the fact that $|\sum W| \leq \sum |W|$, where $W = ((\mathbb{1}\{[\theta]_v^\top(s,a) > 0\} - \mathbb{1}\{[\Theta_{init}]_v^\top(s,a) > 0\})b_v[\theta]_v^\top(s,a))$. Θ_{init} is defined in Eq. (5). Next, since $\mathbb{1}\{[\Theta_{init}]_v^\top(s,a) > 0\} \neq \mathbb{1}\{[\theta]_v^\top(s,a) > 0\}$, we have,

$$|[\Theta_{init}]_v^\top(s,a)| \leq |[\theta]_v^\top(s,a) - \Theta_{init}]_v^\top(s,a)|$$

$$\leq \|[\theta]_v - [\Theta_{init}]_v\|_2, \tag{49}$$

where we obtain the last inequality from the Cauchy-Schwartz inequality. We also assume that $\|(s,a)\|_2 \leq 1$ without loss of generality [31,51]. Equation (49) further implies that,

$$|\mathbb{1}\{[\theta]_v^\top(s,a) > 0\} - \mathbb{1}\{[\Theta_{init}]_v^\top(s,a) > 0\}|$$

$$\leq \mathbb{1}\{|[\Theta_{init}]_v^\top(s,a)| \leq \|[\theta]_v - [\Theta_{init}]_v\|_2\} \tag{50}$$

Then plug Eq. (50) and the fact that $|b_v| \leq 1$ back to Eq. (48), we have the following,

$$\mathbb{E}_{\nu_\pi}\left[\left(f((s,a);\theta) - \hat{f}((s,a);\theta)\right)^2\right]$$

$$\leq \frac{1}{m}\mathbb{E}_{\nu_\pi}\left[\left(\sum_{v=1}^{m}\mathbb{1}\left\{\left|[\Theta_{\text{init}}]_v^\top (s,a)\right| \leq \left\|[\theta]_v - [\Theta_{\text{init}}]_v\right\|_2\right\}\right.\right.$$
$$\left.\left.\left|[\theta]_v^\top (s,a)\right|\right)^2\right]$$

$$\leq \frac{1}{m}\mathbb{E}_{\nu_\pi}\left[\left(\sum_{v=1}^{m}\mathbb{1}\left\{\left|[\Theta_{\text{init}}]_v^\top (s,a)\right| \leq \left\|[\theta]_v - [\Theta_{\text{init}}]_v\right\|_2\right\}\right.\right.$$
$$\left.\left.\left(\left|([\theta]_v - [\Theta_{\text{init}}]_v)^\top (s,a)\right| + \left|[\Theta_{\text{init}}]_v^\top (s,a)\right|\right)\right)^2\right]$$

$$\leq \frac{1}{m}\mathbb{E}_{\nu_\pi}\left[\left(\sum_{v=1}^{m}\mathbb{1}\left\{\left|[\Theta_{\text{init}}]_v^\top (s,a)\right| \leq \left\|[\theta]_v - [\Theta_{\text{init}}]_v\right\|_2\right\}\right.\right.$$
$$\left.\left.\left(\left\|[\theta]_v - [\Theta_{\text{init}}]_v\right\|_2 + \left|[\Theta_{\text{init}}]_v^\top (s,a)\right|\right)\right)^2\right]$$

$$\leq \frac{1}{m}\mathbb{E}_{\nu_\pi}\left[\left(\sum_{v=1}^{m}\mathbb{1}\left\{\left|[\Theta_{\text{init}}]_v^\top (s,a)\right| \leq \left\|[\theta]_v - [\Theta_{\text{init}}]_v\right\|_2\right\}\right.\right.$$
$$\left.\left.2\left\|[\theta]_v - [\Theta_{\text{init}}]_v\right\|_2\right)^2\right] \tag{51}$$

We obtain the second inequality by the fact that $|A| \leq |A - B| + |B|$. Then follow the Cauchy-Schwartz inequality and $\|(s,a)\|_2 \leq 1$ we have the third equality. By inserting Eq. (49) we achieve the fourth inequality. We continue Eq. (51) by following the Cauchy-Schwartz inequality and plugging $\|[\theta] - [\Theta_{\text{init}}]\|_2 \leq \Upsilon$,

$$\mathbb{E}_{\nu_\pi}\left[\left(f((s,a);\theta) - \hat{f}((s,a);\theta)\right)^2\right]$$

$$\leq \frac{4\Upsilon^2}{m}\mathbb{E}_{\nu_\pi}\left[\sum_{v=1}^{m}\mathbb{1}\{|[\Theta_{\text{init}}]_v^\top (s,a)| \leq \|[\theta]_v - [\Theta_{\text{init}}]_v\|_2\}\right]$$

$$= \frac{4\Upsilon^2}{m}\sum_{v=1}^{m}P_{\nu_\pi}|[\Theta_{\text{init}}]_v^\top (s,a)| \leq \|[\theta]_v - [\Theta_{\text{init}}]_v\|_2)$$

$$\leq \frac{4c\Upsilon^2}{m}\sum_{v=1}^{m}\frac{\|[\theta]_v - [\Theta_{\text{init}}]_v\|_2}{\|[\Theta_{\text{init}}]_v\|_2}$$

$$\leq \frac{4c\Upsilon^2}{m}\left(\sum_{v=1}^{m}\|[\theta]_v - [\Theta_{\text{init}}]_v\|_2^2\right)^{-1/2}\left(\sum_{v=1}^{m}\frac{1}{\|[\Theta_{\text{init}}]_v\|_2^2}\right)^{-1/2}$$

$$\leq \frac{4c_1\Upsilon^3}{\sqrt{m}} \tag{52}$$

We obtain the second inequality by imposing Assumption 4 and the third by following the Cauchy-Schwartz inequality. Finally, we set $c_1 := c\sqrt{\mathbb{E}_{\mathcal{N}(0,I_d/d)}[1/\|(s,a)\|_2^2]}$. Thus, we complete the proof.

In the t-th iterations of TD iteration, we denote the temporal difference terms w.r.t $\hat{f}((s,a);\theta_t)$ and $f((s,a);\theta_t)$ as

$$\delta_t^0((s,a),(s,a)';\theta_t) = \hat{f}((s,a)';\theta_t) - \gamma\hat{f}((s,a);\theta_t)$$
$$- r_{s,a},$$
$$\delta_t^\theta((s,a),(s,a)';\theta_t) = f((s,a)';\theta_t) - \gamma f((s,a);\theta_t)$$
$$- r_{s,a}.$$

For notation simplicity in the sequel we write $\delta_t^0((s,a),(s,a)';\theta_t)$ and $\delta_t^\theta((s,a),(s,a)';\theta_t)$ as δ_t^0 and δ_t^θ. We further define the stochastic semi-gradient $g_t(\theta_t) := \delta_t^\theta \nabla_\theta f((s,a);\theta_t)$, its population mean $\bar{g}_t(\theta_t) := \mathbb{E}_{\nu_\pi}[g_t(\theta_t)]$. The local linearization of $\bar{g}_t(\theta_t)$ is $\hat{g}_t(\theta_t) := \mathbb{E}_{\nu_\pi}[\delta_t^0 \nabla_\theta \hat{f}((s,a);\theta_t)]$. We denote them as $g_t, \bar{g}_t, \hat{g}_t$ respectively for simplicity.

Lemma 15. *Under Assumption 4, for all $\theta_t \in \mathcal{D}$, where $\mathcal{D} = \{\xi \in \mathbb{R}^{md} : \|\xi - \Theta_{init}\|_2 \leq \Upsilon\}$, it holds with probability of $1 - \delta$ that,*

$$\|\bar{g}_t - \hat{g}_t\|_2$$
$$= \mathcal{O}\left(\Upsilon^{3/2} m^{-1/4}\left(1 + (m\log\frac{1}{\delta})^{-1/2}\right) + \Upsilon^{1/2} r_{\max} m^{-1/4}\right)$$

Proof. By the definition of $\bar{g}_t$ and $\hat{g}_t$, we have

$$\|\bar{g}_t - \hat{g}_t\|_2^2$$
$$= \left\|\mathbb{E}_{\nu_\pi}[\delta_t^\theta \nabla_\theta f((s,a);\theta_t) - \delta_t^0 \nabla_\theta \hat{f}((s,a);\theta_t)]\right\|_2^2$$
$$= \left\|\mathbb{E}_{\nu_\pi}[(\delta_t^\theta - \delta_t^0)\nabla_\theta f((s,a);\theta_t) + \delta_t^0(\nabla_\theta f((s,a);\theta_t) - \nabla_\theta \hat{f}((s,a);\theta_t))]\right\|_2^2$$
$$\leq 2\mathbb{E}_{\nu_\pi}[(\delta_t^\theta - \delta_t^0)^2 \|\nabla_\theta f((s,a);\theta_t)\|_2^2] +$$
$$2\mathbb{E}_{\nu_\pi}\left[\left(|\delta_t^0|\|\nabla_\theta f((s,a);\theta_t) - \nabla_\theta \hat{f}((s,a);\theta_t))\|_2\right)^2\right] \tag{53}$$

We obtain the inequality because $(A+B)^2 \leq 2A^2 + 2B^2$. *We first upper bound* $\mathbb{E}_{\nu_\pi}[(\delta_t^\theta - \delta_t^0)^2 \|\nabla_\theta f((s,a);\theta_t)\|_2^2]$ *in Eq. (53). Since* $\|(s,a)\|_2 \leq 1$, *we have* $\|\nabla_\theta f((s,a);\theta_t)\|_2 \leq 1$. *Then by definition, we have the following first inequality,*

$$\mathbb{E}_{\nu_\pi}\left[\left(\delta_t^\theta - \delta_t^0\right)^2 \left\|\nabla_\theta f((s,a);\theta_t)\right\|_2^2\right]$$
$$\leq \mathbb{E}_{\nu_\pi}\left[\left(f((s,a);\theta_t) - \hat{f}((s,a);\theta_t) - \gamma\left(f((s',a');\theta_t) - \hat{f}((s',a');\theta_t))\right)\right)^2\right]$$
$$\leq \mathbb{E}_{\nu_\pi}\left[\left(\left|f((s,a);\theta_t) - \hat{f}((s,a);\theta_t)\right| + \left|f((s',a');\theta_t) - \hat{f}((s',a');\theta_t)\right|\right)^2\right]$$

$$\leq 2\mathbb{E}_{\nu_\pi}\left[\left(f((s,a);\theta_t) - \hat{f}((s,a);\theta_t)\right)^2\right] + 2\mathbb{E}_{\nu_\pi}$$
$$\left[\left(f((s',a');\theta_t) - \hat{f}((s',a');\theta_t)\right)^2\right]$$
$$\leq 4\mathbb{E}_{\nu_\pi}\left[\left(f((s,a);\theta_t) - \hat{f}((s,a);\theta_t)\right)^2\right] \leq \frac{16c_1\Upsilon^3}{\sqrt{m}} \tag{54}$$

We obtain the second inequality by $|\gamma| \leq 1$, *then obtain the third inequality by the fact that* $(A+B)^2 \leq 2A^2 + 2B^2$. We reach the final step by inserting Lemma 14. We then proceed to upper bound $\mathbb{E}_{\nu_\pi}\left[|\delta_t^0|\|\nabla_\theta f((s,a);\theta_t) - \nabla_\theta \hat{f}((s,a);\theta_t))\|_2\right]$. From Hölder's inequality, we have,

$$\mathbb{E}_{\nu_\pi}\left[\left(|\delta_t^0|\|\nabla_\theta f((s,a);\theta_t) - \nabla_\theta \hat{f}((s,a);\theta_t))\|_2\right)^2\right]$$
$$\leq \mathbb{E}_{\nu_\pi}\left[(\delta_t^0)^2\right]\mathbb{E}_{\nu_\pi}\left[\|\nabla_\theta f((s,a);\theta_t) - \nabla_\theta \hat{f}((s,a);\theta_t))\|_2^2\right] \tag{55}$$

We first derive an upper bound for first term in Eq. (55), starting from its definition,

$$\mathbb{E}_{\nu_\pi}\left[(\delta_t^0)^2\right]$$
$$= \mathbb{E}_{\nu_\pi}\left[\left[\hat{f}((s',a');\theta_t) - \gamma\hat{f}((s,a);\theta_t) - r_{s,a}\right]^2\right]$$
$$\leq 3\mathbb{E}_{\nu_\pi}\left[\left(\hat{f}((s',a');\theta_t)\right)^2\right] + 3\mathbb{E}_{\nu_\pi}\left[\left(\gamma\hat{f}((s,a);\theta_t)\right)^2\right]$$
$$+ 3\mathbb{E}_{\nu_\pi}\left[r_{s,a}^2\right]$$
$$\leq 6\mathbb{E}_{\nu_\pi}\left[\left(\hat{f}((s,a);\theta_t)\right)^2\right] + 3r_{\max}^2$$
$$= 6\mathbb{E}_{\nu_\pi}\left[\left(\hat{f}((s,a);\theta_t) - \hat{f}((s,a);\theta_{\pi^*}) + \hat{f}((s,a);\theta_{\pi^*})\right.\right.$$
$$\left.\left. - Q_\pi + Q_\pi\right)^2\right] + 3r_{\max}^2$$
$$\leq 18\mathbb{E}_{\nu_\pi}\left[\left(\hat{f}((s,a);\theta_t) - \hat{f}((s,a);\theta_{\pi^*})\right)^2\right] + 18\mathbb{E}_{\nu_\pi}$$
$$\left[\left(\hat{f}((s,a);\theta_{\pi^*}) - Q_\pi\right)^2\right] + 18\mathbb{E}_{\nu_\pi}\left[\left(Q_\pi\right)^2\right] + 3r_{\max}^2$$
$$\leq 72\Upsilon^2 + 18\mathbb{E}_{\nu_\pi}\left[\left(\hat{f}((s,a);\theta_{\pi^*}) - Q_\pi\right)^2\right]$$
$$+ 21(1-\gamma)^{-2}r_{\max}^2 \tag{56}$$

We obtain the first and the third inequality by the fact that $(A + B + C)^2 \leq 3A^2 + 3B^2 + 3C^2$. Recall $r_{\max}$ is the boundary for reward function r, which leads to the second inequality. We obtain the last inequality in Eq. (56) following the fact that $|\hat{f}((s,a);\theta_t) - \hat{f}((s,a);\theta_{\pi^*})| \leq \|\theta_t - \theta_{\pi^*}\| \leq 2\Upsilon$ and $Q_\pi \leq (1-\gamma)^{-1}r_{\max}$. Since $\bar{\mathcal{F}}_{\Upsilon,m} \subset \mathcal{F}_{\Upsilon,m}$, by Lemma 13, we have,

$$E_{\nu_\pi}\left[\left(\hat{f}((s,a);\theta_{\pi^*}) - Q_\pi\right)^2\right] \leq \frac{\Upsilon^2\left(1 + \sqrt{2\log(1/\delta)}\right)^2}{m} \tag{57}$$

Combine Eq. (56) and Eq. (57), we have with probability of $1 - \delta$,

$$\mathbb{E}_{\nu_\pi}\left[(\delta_t^0)^2\right]$$

$$\leq 72\Upsilon^2(1 + \frac{\log(1/\delta)}{m}) + 21(1-\gamma)^{-2}r_{\max}^2 \tag{58}$$

Lastly we have

$$\mathbb{E}_{\nu_\pi}\left[\|\nabla_\theta f((s,a);\theta_t) - \nabla_\theta \hat{f}((s,a);\theta_t))\|_2^2\right]$$

$$= \mathbb{E}_{\nu_\pi}\left[\left(\frac{1}{m}\sum_{v=1}^m \left(\mathbb{1}\{[\theta]_v^\top(s,a) > 0\} - \mathbb{1}\{[\Theta_{\text{init}}]_v^\top(s,a)\right.\right.\right.$$

$$\left.\left.\left.> 0\}\right)^2(b_v)^2\|(s,a)\|_2^2\right)\right]$$

$$\leq \mathbb{E}_{\nu_\pi}\left[\frac{1}{m}\sum_{v=1}^m \left(\mathbb{1}\{|[\Theta_{\text{init}}]_v^\top(s,a)| \leq \|[\theta]_v - [\Theta_{\text{init}}]_v\|_2\}\right)\right]$$

$$\leq \frac{c_1\Upsilon}{\sqrt{m}} \tag{59}$$

We obtain the first inequality by following Eq. (50) and the fact that $|b_v| \leq 1$ and $\|(s,a)\|_2 \leq 1$. Then for the rest, we follow the similar argument in Eq. (52). To finish the proof, we plug Eq. (54), Eq. (58) and Eq. (59) back to Eq. (53),

$$\|\bar{g}_t - \hat{g}_t\|_2^2$$

$$\leq 2\left(\frac{16c_1\Upsilon^3}{\sqrt{m}} + \left(72\Upsilon^2(1 + \frac{\log(1/\delta)}{m}) + 21(1-\gamma)^{-2}r_{\max}^2\right)\frac{c_1\Upsilon}{\sqrt{m}}\right)$$

$$= \frac{176c_1\Upsilon^3}{\sqrt{m}} + \frac{144c_1\Upsilon^3\log(1/\delta)}{m^{3/2}} + \frac{42c_1\Upsilon r_{\max}^2}{(1-\gamma)^{-2}\sqrt{m}}$$

Then we have,

$$\|\bar{g}_t - \hat{g}_t\|_2$$

$$\leq \sqrt{\frac{176c_1\Upsilon^3}{\sqrt{m}} + \frac{144c_1\Upsilon^3\log(1/\delta)}{m^{3/2}} + \frac{42c_1\Upsilon r_{\max}^2}{(1-\gamma)^{-2}\sqrt{m}}}$$

$$\leq \sqrt{\frac{176c_1\Upsilon^3}{\sqrt{m}}} + \sqrt{\frac{144c_1\Upsilon^3\log(1/\delta)}{m^{3/2}}} + \sqrt{\frac{42c_1\Upsilon r_{\max}^2}{(1-\gamma)^{-2}\sqrt{m}}}$$

$$= \mathcal{O}\left(\Upsilon^{3/2}m^{-1/4}(1 + (m\log\frac{1}{\delta})^{-1/2}) + \Upsilon^{1/2}r_{\max}m^{-1/4}\right)$$

Next, we provide the following lemma to characterize the variance of g_t.

Lemma 16 *(Variance of the Stochastic Update Vector)* [31]. *There exists a constant $\xi_g^2 = \mathcal{O}(\Upsilon^2)$ independent of t. Such that for any $t \leq T$, it holds that*

$$\mathbb{E}_{\nu_\pi}[\|g_t(\theta_t) - \bar{g}_t(\theta_t)\|_2^2] \leq \xi_g^2$$

A detailed proof can be found in [31]. Now we provide the proof for Lemma 2.

Proof.

$$
\begin{aligned}
&\left\|\theta_{t+1} - \theta_{\pi^*}\right\|_2^2 \\
&= \left\|\Pi_{\mathcal{D}}(\theta_t - \eta g_t(\theta_t)) - \Pi_{\mathcal{D}}(\theta_{\pi^*} - \eta \hat{g}_t(\theta_{\pi^*}))\right\|_2^2 \\
&\leq \left\|(\theta_t - \theta_{\pi^*}) - \eta(g_t(\theta_t) - \hat{g}_t(\theta_{\pi^*}))\right\|_2^2 \\
&= \left\|\theta_t - \theta_{\pi^*}\right\|_2^2 - 2\eta\big(g_t(\theta_t) - \hat{g}_t(\theta_{\pi^*})\big)^{\top}(\theta_t - \theta_{\pi^*}) \\
&\quad + \eta^2\left\|g_t(\theta_t) - \hat{g}_t(\theta_{\pi^*})\right\|_2^2
\end{aligned}
\tag{60}
$$

The inequality holds due to the definition of $\Pi_{\mathcal{D}}$. We first upper bound $\left\|g_t(\theta_t) - \hat{g}_t(\theta_{\pi^*})\right\|_2^2$ in Eq. (60),

$$
\begin{aligned}
&\left\|g_t(\theta_t) - \hat{g}_t(\theta_{\pi^*})\right\|_2^2 \\
&= \left\|g_t(\theta_t) - \bar{g}_t(\theta_t) + \bar{g}_t(\theta_t) - \hat{g}_t(\theta_t) + \hat{g}_t(\theta_t) - \hat{g}_t(\theta_{\pi^*})\right\|_2^2 \\
&\leq 3\Big(\left\|g_t(\theta_t) - \bar{g}_t(\theta_t)\right\|_2^2 + \left\|\bar{g}_t(\theta_t) - \hat{g}_t(\theta_t)\right\|_2^2 + \\
&\quad \left\|\hat{g}_t(\theta_t) - \hat{g}_t(\theta_{\pi^*})\right\|_2^2\Big)
\end{aligned}
\tag{61}
$$

The inequality holds due to fact that $(A + B + C)^2 \leq 3A^2 + 3B^2 + 3C^2$. Two of the terms on the right hand side of Eq. (61) are characterized in Lemma 15 and Lemma 16. We therefore characterize the remaining term,

$$
\begin{aligned}
&\left\|\hat{g}_t(\theta_t) - \hat{g}_t(\theta_{\pi^*})\right\|_2^2 \\
&= \mathbb{E}_{\nu_\pi}\Big[\big(\delta_t^0(\theta_t) - \delta_t^0(\theta_{\pi^*})\big)^2 \left\|\nabla_\theta \hat{f}((s,a);\theta_t)\right\|_2^2\Big] \\
&\leq \mathbb{E}_{\nu_\pi}\Big[\Big(\big(\hat{f}((s,a);\theta_t) - \hat{f}((s,a);\theta_{\pi^*})\big) - \gamma\big(\hat{f}((s',a'); \\
&\quad \theta_t) - \hat{f}((s',a');\theta_{\pi^*})\big)\Big)^2\Big] \\
&\leq \mathbb{E}_{\nu_\pi}\Big[\big(\hat{f}((s,a);\theta_t) - \hat{f}((s,a);\theta_{\pi^*})\big)^2\Big] + 2\gamma\mathbb{E}_{\nu_\pi} \\
&\quad \Big[\big(\hat{f}((s',a');\theta_t) - \hat{f}((s',a');\theta_{\pi^*})\big)\big(\hat{f}((s,a);\theta_t) \\
&\quad - \hat{f}((s,a);\theta_{\pi^*})\big)\Big] \\
&\quad + \gamma^2\mathbb{E}_{\nu_\pi}\Big[\big(\hat{f}((s',a');\theta_t) - \hat{f}((s',a');\theta_{\pi^*})\big)^2\Big]
\end{aligned}
\tag{62}
$$

We obtain the first inequality by the fact that $\left\|\nabla_\theta \hat{f}((s,a);\theta_t)\right\|_2 \leq 1$. Then we use the fact that (s,a) and (s',a') have the same marginal distribution as well as $\gamma < 1$ for the second inequality. Follow the Cauchy-Schwarz inequality and the fact that (s, a) and (s', a') have the same marginal distribution, we have

$$\mathbb{E}_{\nu_\pi}\left[\left(\hat{f}((s', a'); \theta_t) - \hat{f}((s', a'); \theta_{\pi^*})\right)\left(\hat{f}((s, a); \theta_t)\right.\right.$$
$$\left.\left. - \hat{f}((s, a); \theta_{\pi^*})\right)\right]$$
$$\leq \mathbb{E}_{\nu_\pi}\left[\left(\hat{f}((s', a'); \theta_t) - \hat{f}((s', a'); \theta_{\pi^*})\right)\right]\mathbb{E}_{\nu_\pi}$$
$$\left[\left(\hat{f}((s, a); \theta_t) - \hat{f}((s, a); \theta_{\pi^*})\right)\right]$$
$$= \mathbb{E}_{\nu_\pi}\left[\left(\hat{f}((s', a'); \theta_t) - \hat{f}((s', a'); \theta_{\pi^*})\right)^2\right] \tag{63}$$

We plug Eq. (63) back to Eq. (62),

$$\left\|\hat{g}_t(\theta_t) - \hat{g}_t(\theta_{\pi^*})\right\|_2^2$$
$$\leq (1 + \gamma)^2 \mathbb{E}_{\nu_\pi}\left[\left(\hat{f}((s, a); \theta_t) - \hat{f}((s, a); \theta_{\pi^*})\right)^2\right]. \tag{64}$$

Next, we upper bound $\left(g_t(\theta_t) - \hat{g}_t(\theta_{\pi^*})\right)^\top (\theta_t - \theta_{\pi^*})$. We have,

$$\left(g_t(\theta_t) - \hat{g}_t(\theta_{\pi^*})\right)^\top (\theta_t - \theta_{\pi^*})$$
$$= \left(g_t(\theta_t) - \bar{g}_t(\theta_t)\right)^\top (\theta_t - \theta_{\pi^*}) + \left(\bar{g}_t(\theta_t) - \hat{g}_t(\theta_t)\right)^\top$$
$$(\theta_t - \theta_{\pi^*}) + \left(\hat{g}_t(\theta_t) - \hat{g}_t(\theta_{\pi^*})\right)^\top (\theta_t - \theta_{\pi^*}) \tag{65}$$

One term on the right hand side of Eq. (65) are characterized by Lemma 16. We continue to characterize the remaining terms. First, by Hölder's inequality, we have

$$\left(\bar{g}_t(\theta_t) - \hat{g}_t(\theta_t)\right)^\top (\theta_t - \theta_{\pi^*})$$
$$\geq -\left\|\bar{g}_t(\theta_t) - \hat{g}_t(\theta_t)\right\|_2 \|\theta_t - \theta_{\pi^*}\|_2$$
$$\geq -2\Upsilon\left\|\bar{g}_t(\theta_t) - \hat{g}_t(\theta_t)\right\|_2 \tag{66}$$

We obtain the second inequality since $\|\theta_t - \theta_{\pi^*}\|_2 \leq 2\Upsilon$ by definition. For the last term,

$$\left(\hat{g}_t(\theta_t) - \hat{g}_t(\theta_{\pi^*})\right)^\top (\theta_t - \theta_{\pi^*})$$
$$= \mathbb{E}_{\nu_\pi}\left[\left(\left(\hat{f}((s, a); \theta_t) - \hat{f}((s, a); \theta_{\pi^*})\right) - \gamma\left(\hat{f}((s', a'); \theta_t)\right.\right.\right.$$
$$\left.\left.\left. - \hat{f}((s', a'); \theta_{\pi^*})\right)\right)\left(\nabla_\theta \hat{f}((s, a); \theta_t)\right)^\top (\theta_t - \theta_{\pi^*})\right]$$
$$= \mathbb{E}_{\nu_\pi}\left[\left(\left(\hat{f}((s, a); \theta_t) - \hat{f}((s, a); \theta_{\pi^*})\right) - \gamma\left(\hat{f}((s', a'); \theta_t)\right.\right.\right.$$
$$\left.\left.\left. - \hat{f}((s', a'); \theta_{\pi^*})\right)\right)\left(\hat{f}((s, a); \theta_t) - \hat{f}((s, a); \theta_{\pi^*})\right)\right]$$

$$\geq \mathbb{E}_{\nu_\pi}\left[\left(\left(\hat{f}((s,a);\theta_t) - \hat{f}((s,a);\theta_{\pi^*})\right)\right)^2\right]$$

$$- \gamma\mathbb{E}_{\nu_\pi}\left[\left(\left(\hat{f}((s,a);\theta_t) - \hat{f}((s,a);\theta_{\pi^*})\right)\right)^2\right]$$

$$= (1-\gamma)\mathbb{E}_{\nu_\pi}\left[\left(\hat{f}((s,a);\theta_t) - \hat{f}((s,a);\theta_{\pi^*})\right)^2\right], \tag{67}$$

where the inequality follows from Eq. (63). Combine Eqs. (60), (61), (64), (65), (66) and (67), we have,

$$\|\theta_{t+1} - \theta_{\pi^*}\|_2^2$$
$$\leq \|\theta_t - \theta_{\pi^*}\|_2^2 - \left(2\eta(1-\gamma) - 3\eta^2(1+\gamma)^2\right)$$
$$\mathbb{E}_{\nu_\pi}\left[\left(\hat{f}((s,a);\theta_t) - \hat{f}((s,a);\theta_{\pi^*})\right)^2\right]$$
$$+ 3\eta^2\|\bar{g}_t - \hat{g}_t\|_2^2 + 4\eta\Upsilon\|\bar{g}_t - \hat{g}_t\|_2 + 4\Upsilon\eta|\xi_g|$$
$$+ 3\eta^2\xi_g^2 \tag{68}$$

We then bound the error terms by rearrange Eq. (68). First, we have, with probability of $1 - \delta$,

$$\mathbb{E}_{\nu_\pi}\left[\left(f((s,a);\theta_t) - \hat{f}((s,a);\theta_{\pi^*})\right)^2\right]$$
$$= \mathbb{E}_{\nu_\pi}\left[\left(f((s,a);\theta_t) - \hat{f}((s,a);\theta_t) + \hat{f}((s,a);\theta_t)\right.\right.$$
$$\left.\left. - \hat{f}((s,a);\theta_{\pi^*})\right)^2\right]$$
$$\leq 2\mathbb{E}_{\nu_\pi}\left[\left(f((s,a);\theta_t) - \hat{f}((s,a);\theta_t)\right)^2 + \left(\hat{f}((s,a);\theta_t)\right.\right.$$
$$\left.\left. - \hat{f}((s,a);\theta_{\pi^*})\right)^2\right]$$
$$\leq \left(\eta(1-\gamma) - 1.5\eta^2(1+\gamma)^2\right)^{-1}\left(\|\theta_t - \theta_{\pi^*}\|_2^2\right.$$
$$\left. - \|\theta_{t+1} - \theta_{\pi^*}\|_2^2 + 4\Upsilon\eta|\xi_g| + 3\eta^2\xi_g^2\right) + \epsilon_g \tag{69}$$

where

$$\epsilon_g = \mathcal{O}(\Upsilon^3 m^{-1/2}\log(1/\delta) + \Upsilon^{5/2}m^{-1/4}\sqrt{\log(1/\delta)}$$
$$+ \Upsilon r_{\max}^2 m^{-1/4})$$

We obtain the first inequality by the fact that $(A+B)^2 \leq 2A^2 + 2B^2$. Then by Eq. (68), Lemma 14 and Lemma 15, we reach the final inequality. By telescoping Eq. (69) for $t =$ to T, we have, with probability of $1 - \delta$,

$$\left\| f\big((s,a);\theta_T\big) - \hat{f}\big((s,a);\theta_{\pi^*}\big) \right\|^2$$

$$\leq \frac{1}{T}\sum_{t=1}^{T}\mathbb{E}_{\nu_\pi}\left[\Big(f\big((s,a);\theta_t\big) - \hat{f}\big((s,a);\theta_{\pi^*}\big)\Big)^2\right]$$

$$\leq T^{-1}\big(2\eta(1-\gamma) - 3\eta^2(1+\gamma)^2\big)^{-1}(\|\Theta_{\text{init}} - \theta_{\pi^*}\| +$$
$$4\Upsilon T\eta|\xi_g| + 3T\eta^2\xi_g^2) + \epsilon_g$$

Set $\eta = \min\{1/\sqrt{T}, (1-\gamma)/3(1+\gamma)^2\}$, which implies that $T^{-1/2}(2\eta(1-\gamma) - 3\eta^2(1+\gamma)^2)^{-1} \leq 1/(1-\gamma)^2$, then we have, with probability of $1-\delta$,

$$\left\| f\big((s,a);\theta_T\big) - \hat{f}\big((s,a);\theta_{\pi^*}\big) \right\|$$

$$\leq \frac{1}{(1-\gamma)^2\sqrt{T}}(\|\Theta_{\text{init}} - \theta_{\pi^*}\|_2^2 + 4\Upsilon\sqrt{T}|\xi_g|$$
$$+ 3\xi_g^2) + \epsilon_g$$

$$\leq \frac{\Upsilon^2 + 4\Upsilon\sqrt{T}|\xi_g| + 3\xi_g^2}{(1-\gamma)^2\sqrt{T}} + \epsilon_g$$

$$= \mathcal{O}(\Upsilon^3 m^{-1/2}\log(1/\delta) + \Upsilon^{5/2}m^{-1/4}\sqrt{\log(1/\delta)}$$
$$+ \Upsilon r_{\max}^2 m^{-1/4} + \Upsilon^2 T^{-1/2} + \Upsilon)$$

We obtain the second inequality by the fact that $\|\Theta_{\text{init}} - \theta_{\pi^*}\|_2 \leq \Upsilon$. Then by definition we replace $\tilde{Q}_{\omega_k}$ and $\tilde{Q}_{\pi_k}$

E Additional Related Work

E.1 Global Optimality of Policy Search Methods

A major challenge of existing RL research is the lack of theoretical justification, such as sample complexity analysis, mainly because the objective function of policy search in RL is often nonconvex. It is challenging to determine if a policy search approach is guaranteed to reach the global optimal. Besides, the RL architecture components are usually parameterized by neural networks in practice. Its nonlinearity and complex nature render the analysis significantly difficult [62].

The theoretical understanding of policy gradient methods is also under tentative study. Work on this topic has been done mostly in tabular and linear parametrization settings for different variants of policy gradient. For example, [11] and [44] establish a non-asymptotic convergence guarantee for natural policy gradient (NPG, [22]) and trusted region policy optimization (TRPO, [42]), respectively. [35] show converge rate for softmax parametrization, while [1] analyze multiple variants of policy gradient. On the other side of the spectrum, [31,51] prove the global convergence and optimality of various policy gradient algorithms with over-parameterized neural networks. Furthermore, [62] apply the global optimality analysis to variance-constrained actor-critic risk-averse control with cumulative average rewards, and proposed a corresponding variance-constrained actor-critic (VARAC) algorithm. However, the analysis procedure is complicated due to the risk constraints on cumulative rewards, and the algorithm's experimental performance remains unverified. Therefore, it remains interesting if there can be simplified global optimality analysis with verifiable experimental studies for risk-averse policy search methods.

E.2   Over-Parameterized Neural Networks in RL

Overparameterization, a technique of deploying more parameters than necessary, improves the performance of neural networks [59]. The learning ability and generalization of over-parameterized neural networks have been studied extensively [2,5,15]. Integration with over-parameterized neural networks can be found in multiple RL topics. One line of work is to prove the global optimality of RL algorithms in a non-linear approximation setting [31,51,62]. They use ReLU activation over-parameterized neural networks with policy gradient methods such as NPG and PPO. Our work also belongs to this category. Other works include [19], which also deploy a two-layered ReLU activation over-parameterized neural network on mean-field multi-agent reinforcement learning problem. Regularization with over-parameterized neural networks is also investigated recently [25,41].

References

1. Agarwal, A., Kakade, S.M., Lee, J.D., Mahajan, G.: On the theory of policy gradient methods: optimality, approximation, and distribution shift. J. Mach. Learn. Res. **22**(98), 1–76 (2021)
2. Allen-Zhu, Z., Li, Y., Liang, Y.: Learning and generalization in overparameterized neural networks, going beyond two layers. In: Advances in Neural Information Processing Systems 32 (2019)
3. Allen-Zhu, Z., Li, Y., Song, Z.: A convergence theory for deep learning via over-parameterization (2019)
4. Antos, A., Szepesvári, C., Munos, R.: Fitted Q-iteration in continuous action-space MDPs. In: Advances in Neural Information Processing Systems 20 (2007)
5. Arora, S., Du, S., Hu, W., Li, Z., Wang, R.: Fine-grained analysis of optimization and generalization for overparameterized two-layer neural networks. In: International Conference on Machine Learning, pp. 322–332. PMLR (2019)
6. Bhandari, J., Russo, D.: Global optimality guarantees for policy gradient methods. arXiv preprint arXiv:1906.01786 (2019)
7. Bisi, L., Sabbioni, L., Vittori, E., Papini, M., Restelli, M.: Risk-averse trust region optimization for reward-volatility reduction. In: Bessiere, C. (ed.) Proceedings of the Twenty-Ninth International Joint Conference on Artificial Intelligence, IJCAI-2020, pp. 4583–4589. International Joint Conferences on Artificial Intelligence Organization, July 2020. Special Track on AI in FinTech
8. Brockman, G., et al.: OpenAI gym. arXiv preprint arXiv:1606.01540 (2016)
9. Cai, Q., Yang, Z., Lee, J.D., Wang, Z.: Neural temporal-difference and Q-learning provably converge to global optima. arXiv preprint arXiv:1905.10027 (2019)

10. Cao, Y., Gu, Q.: Generalization error bounds of gradient descent for learning over-parameterized deep ReLU networks. In: Proceedings of the AAAI Conference on Artificial Intelligence, vol. 34, pp. 3349–3356 (2020)
11. Cen, S., Cheng, C., Chen, Y., Wei, Y., Chi, Y.: Fast global convergence of natural policy gradient methods with entropy regularization. Oper. Res. **70**(4), 2563–2578 (2021)
12. Csiszár, I., Körner, J.: Information Theory: Coding Theorems for Discrete Memoryless Systems. Cambridge University Press, Cambridge (2011)
13. Dabney, W., et al.: A distributional code for value in dopamine-based reinforcement learning. Nature **577**(7792), 671–675 (2020)
14. Di Castro, D., Tamar, A., Mannor, S.: Policy gradients with variance related risk criteria. arXiv preprint arXiv:1206.6404 (2012)
15. Du, S.S., Zhai, X., Poczos, B., Singh, A.: Gradient descent provably optimizes over-parameterized neural networks. arXiv preprint arXiv:1810.02054 (2018)
16. Farahmand, A.M., Ghavamzadeh, M., Szepesvári, C., Mannor, S.: Regularized policy iteration with nonparametric function spaces. J. Mach. Learn. Res. **17**(1), 4809–4874 (2016)
17. Fu, Z., Yang, Z., Wang, Z.: Single-timescale actor-critic provably finds globally optimal policy. arXiv preprint arXiv:2008.00483 (2020)
18. García, J., Fernández, F.: A comprehensive survey on safe reinforcement learning. J. Mach. Learn. Res. **16**(1), 1437–1480 (2015)
19. Gu, H., Guo, X., Wei, X., Xu, R.: Mean-field multi-agent reinforcement learning: a decentralized network approach. arXiv preprint arXiv:2108.02731 (2021)
20. Hans, A., Schneegaß, D., Schäfer, A.M., Udluft, S.: Safe exploration for reinforcement learning. In: ESANN, pp. 143–148. Citeseer (2008)
21. Kakade, S., Langford, J.: Approximately optimal approximate reinforcement learning. In: In Proceedings of the 19th International Conference on Machine Learning. Citeseer (2002)
22. Kakade, S.M.: A natural policy gradient. In: Advances in Neural Information Processing Systems 14 (2001)
23. Konstantopoulos, T., Zerakidze, Z., Sokhadze, G.: Radon-Nikodým theorem. In: Lovric, M. (ed.) International Encyclopedia of Statistical Science, pp. 1161–1164. Springer, Heidelberg (2011). https://doi.org/10.1007/978-3-642-04898-2_468
24. Kovács, B.: Safe reinforcement learning in long-horizon partially observable environments (2020)
25. Kubo, M., Banno, R., Manabe, H., Minoji, M.: Implicit regularization in over-parameterized neural networks. arXiv preprint arXiv:1903.01997 (2019)
26. La, P., Ghavamzadeh, M.: Actor-critic algorithms for risk-sensitive MDPs. In: Burges, C.J.C., Bottou, L., Welling, M., Ghahramani, Z., Weinberger, K.Q. (eds.) Advances in Neural Information Processing Systems, vol. 26. Curran Associates, Inc. (2013)
27. Lai, T.L., Xing, H., Chen, Z.: Mean-variance portfolio optimization when means and covariances are unknown. Ann. Appl. Stat. **5**(2A), June 2011. https://doi.org/10.1214/10-aoas422
28. Laroche, R., Tachet des Combes, R.: Dr Jekyll and Mr Hyde: the strange case of off-policy policy updates. In: Advances in Neural Information Processing Systems 34 (2021)
29. Li, D., Ng, W.L.: Optimal dynamic portfolio selection: multiperiod mean-variance formulation. Math. Financ. **10**(3), 387–406 (2000)

30. Liu, B., Liu, J., Ghavamzadeh, M., Mahadevan, S., Petrik, M.: Finite-sample analysis of proximal gradient TD algorithms. In: Proceedings of the Conference on Uncertainty in AI (UAI), pp. 504–513 (2015)

31. Liu, B., Cai, Q., Yang, Z., Wang, Z.: Neural trust region/proximal policy optimization attains globally optimal policy. In: Advances in Neural Information Processing Systems 32 (2019)

32. Majumdar, A., Pavone, M.: How should a robot assess risk? Towards an axiomatic theory of risk in robotics. In: Amato, N.M., Hager, G., Thomas, S., Torres-Torriti, M. (eds.) Robotics Research. SPAR, vol. 10, pp. 75–84. Springer, Cham (2020). https://doi.org/10.1007/978-3-030-28619-4_10

33. Mannor, S., Tsitsiklis, J.: Mean-variance optimization in Markov decision processes. arXiv preprint arXiv:1104.5601 (2011)

34. Markowitz, H.M., Todd, G.P.: Mean-Variance Analysis in Portfolio Choice and Capital Markets, vol. 66. Wiley, New York (2000)

35. Mei, J., Xiao, C., Szepesvari, C., Schuurmans, D.: On the global convergence rates of softmax policy gradient methods. In: International Conference on Machine Learning, pp. 6820–6829. PMLR (2020)

36. Mnih, V., et al.: Human-level control through deep reinforcement learning. Nature **518**(7540), 529–533 (2015)

37. Munos, R.: Performance bounds in Lp-norm for approximate value iteration. SIAM J. Control. Optim. **46**(2), 541–561 (2007)

38. Munos, R., Szepesvári, C.: Finite-time bounds for fitted value iteration. J. Mach. Learn. Res. **9**(5), 815–857 (2008)

39. Parker, D.: Managing risk in healthcare: understanding your safety culture using the Manchester patient safety framework (MaPSaF). J. Nurs. Manag. **17**(2), 218–222 (2009)

40. Rahimi, A., Recht, B.: Weighted sums of random kitchen sinks: replacing minimization with randomization in learning. In: Advances in Neural Information Processing Systems 21 (2008)

41. Satpathi, S., Gupta, H., Liang, S., Srikant, R.: The role of regularization in overparameterized neural networks. In: 2020 59th IEEE Conference on Decision and Control (CDC), pp. 4683–4688. IEEE (2020)

42. Schulman, J., Levine, S., Abbeel, P., Jordan, M., Moritz, P.: Trust region policy optimization. In: International Conference on Machine Learning, pp. 1889–1897. PMLR (2015)

43. Schulman, J., Wolski, F., Dhariwal, P., Radford, A., Klimov, O.: Proximal policy optimization algorithms. arXiv preprint arXiv:1707.06347 (2017)

44. Shani, L., Efroni, Y., Mannor, S.: Adaptive trust region policy optimization: global convergence and faster rates for regularized MDPs. In: Proceedings of the AAAI Conference on Artificial Intelligence, vol. 34, pp. 5668–5675 (2020)

45. Sobel, M.J.: The variance of discounted Markov decision processes. J. Appl. Probab. **19**(4), 794–802 (1982)

46. Sutton, R.S., Barto, A.G.: Reinforcement Learning: An Introduction. A Bradford Book. MIT Press, Cambridge (2018)

47. Sutton, R.S., et al.: Fast gradient-descent methods for temporal-difference learning with linear function approximation. In: International Conference on Machine Learning, pp. 993–1000 (2009)

48. Thomas, G., Luo, Y., Ma, T.: Safe reinforcement learning by imagining the near future. In: Advances in Neural Information Processing Systems 34 (2021)

49. Todorov, E., Erez, T., Tassa, Y.: MuJoCo: a physics engine for model-based control. In: 2012 IEEE/RSJ International Conference on Intelligent Robots and Systems, pp. 5026–5033. IEEE (2012)
50. Vinyals, O., et al.: Grandmaster level in StarCraft II using multi-agent reinforcement learning. Nature **575**(7782), 350–354 (2019)
51. Wang, L., Cai, Q., Yang, Z., Wang, Z.: Neural policy gradient methods: global optimality and rates of convergence (2019)
52. Wang, M., Fang, E.X., Liu, H.: Stochastic compositional gradient descent: algorithms for minimizing compositions of expected-value functions. Math. Program. **161**(1–2), 419–449 (2017)
53. Wang, W.Y., Li, J., He, X.: Deep reinforcement learning for NLP. In: Proceedings of the 56th Annual Meeting of the Association for Computational Linguistics: Tutorial Abstracts, pp. 19–21 (2018)
54. Weng, J., Duburcq, A., You, K., Chen, H.: MuJoCo benchmark (2020). https://tianshou.readthedocs.io/en/master/tutorials/benchmark.html
55. Xie, T., et al.: A block coordinate ascent algorithm for mean-variance optimization. In: Bengio, S., Wallach, H., Larochelle, H., Grauman, K., Cesa-Bianchi, N., Garnett, R. (eds.) Advances in Neural Information Processing Systems, vol. 31. Curran Associates, Inc. (2018). https://proceedings.neurips.cc/paper/2018/file/4e4b5fbbbb602b6d35bea8460aa8f8e5-Paper.pdf
56. Xu, P., Chen, J., Zou, D., Gu, Q.: Global convergence of Langevin dynamics based algorithms for nonconvex optimization. In: Advances in Neural Information Processing Systems (2018)
57. Xu, T., Liang, Y., Lan, G.: CRPO: a new approach for safe reinforcement learning with convergence guarantee. In: International Conference on Machine Learning, pp. 11480–11491. PMLR (2021)
58. Yang, L., Wang, M.: Reinforcement learning in feature space: matrix bandit, kernels, and regret bound. In: International Conference on Machine Learning, pp. 10746–10756. PMLR (2020)
59. Zhang, C., Bengio, S., Hardt, M., Recht, B., Vinyals, O.: Understanding deep learning (still) requires rethinking generalization. Commun. ACM **64**(3), 107–115 (2021)
60. Zhang, S., Liu, B., Whiteson, S.: Mean-variance policy iteration for risk-averse reinforcement learning. In: AAAI Conference on Artificial Intelligence (AAAI) (2021)
61. Zhang, S., Tachet, R., Laroche, R.: Global optimality and finite sample analysis of softmax off-policy actor critic under state distribution mismatch. arXiv preprint arXiv:2111.02997 (2021)
62. Zhong, H., Fang, E.X., Yang, Z., Wang, Z.: Risk-sensitive deep RL: variance-constrained actor-critic provably finds globally optimal policy (2020)
63. Zou, D., Cao, Y., Zhou, D., Gu, Q.: Gradient descent optimizes over-parameterized deep ReLU networks. Mach. Learn. **109**(3), 467–492 (2020)

School's Out? Simulating Schooling Strategies During COVID-19

Lukas Tapp[✉], Veronika Kurchyna[✉], Falco Nogatz, Jan Ole Berndt,
and Ingo J. Timm

Smart Data & Knowledge Services, Cognitive Social Simulation, German Research
Center for Artificial Intelligence (DFKI), Trier, Germany
{lukas.tapp,veronika.kurchyna,falco.nogatz,jan_ole.berndt,
ingo.timm}@dfki.de

Abstract. Multi-agent based systems offer the possibility to examine
the effects of policies down to specific target groups while also considering the effects on a population-level scale. To examine the impact of
different schooling strategies, an agent-based model is used in the context
of the COVID-19 pandemic using a German city as an example. The simulation experiments show that reducing the class size by rotating weekly
between in-person classes and online schooling is effective at preventing
infections while driving up the detection rate among children through
testing during weeks of in-person attendance. While open schools lead
to higher infection rates, a surprising result of this study is that school
rotation is almost as effective at lowering infections among both the student population and the general population as closing schools. Due to the
continued testing of attending students, the overall infections in the general population are even lower in a school rotation scenario, showcasing
the potential for emergent behaviors in agent-based models.

Keywords: COVID-19 simulation · Non-pharmaceutical intervention ·
Policy-making and evaluation

1 Introduction

Since the beginning of the COVID-19 pandemic in early 2020, policymakers
across the globe face a novel virus spreading at an unprecedented scale. Without
experience to rely on, governments often struggle to contain the spread of the
virus. Quickly, a flood of data and information became available to decision
makers on all levels of government. Infection rates in districts and counties,
unemployment statistics, the current strain on health systems and critical care
facilities, the financial impact of lockdowns and strict hygiene measures, social
media – a variety of input that must be considered when making decisions.

The researchers have advised policymakers in various German crisis response
groups using a novel dashboard, which approaches the current issues decision
makers face from two angles: The dashboard offers a compact overview of important data from various sources, allowing policymakers to gain a faster understanding of the current situation. Additionally, the dashboard is connected to

© Springer Nature Switzerland AG 2022
F. S. Melo and F. Fang (Eds.): AAMAS 2022 Workshops, LNAI 13441, pp. 48–59, 2022.
https://doi.org/10.1007/978-3-031-20179-0_2

the agent-based *SoSAD model* (Social Simulation for Analysis of Infectious Disease Control) [18]. In this model, the inhabitants of a city are modeled as agents who follow their daily schedules and may spread the disease during interactions. The simulation model enables users to examine the anticipated effects of different non-pharmaceutical interventions such as mask mandates, mandated home office for workers, closing schools, or other measures that aim at reducing infectious contacts.

Analysing different strategies that allow handling the pandemic in schools is the main objective of the paper, based on the counseling work done in different crisis response groups. Since the closing of schools has a strong negative impact on the psychological and intellectual development of children [10], it is important to examine how to keep the number of students in schools at a high level while simultaneously avoiding high disease rates among students and its impact on the general population.

This paper discusses the modeling of infectious diseases with particular focus on uses for policy-making in Sect. 2 before presenting the approach of the SoSAD model in Sect. 3 and how it was used to examine different schooling strategies in Sect. 4. First promising simulation results are presented in Sect. 5, followed by an evaluation of the model itself in Sect. 6. Finally, in Sect. 7, we discuss future work and conclude. After all, this work also explains why this use case is a prime example of the usefulness of agent-based models (ABM) in policy-making contexts.

2   Agent-Based Models in the Pandemic

To predict future behavior in context of pandemics, different simulation studies were conducted since the beginning of the pandemic [11]. Most used a traditional mathematical macro-scale approach [16]. However, many were not capable of simulating social and behavioral factors, such as individual response to countermeasures or social relationships like families living together in a household [17]. ABMs are better suited to express the complexity between individuals. Within multi-agent based systems, many approaches choose a network model in which diseases spread along connections between agents, centering the simulation around relationships. However, this approach doesn't consider that infection chains are often hard to trace [3], as people don't have a static set of people to interact with. Further, such network models have a reduced capacity for implementing individual measures that are specific to certain locations, such as vaccine mandates at workplaces, reduced contact rates, and the closing of schools.

While policymakers have no access to the decision-making and relationships of people, they can influence the behavior of people by setting rules and limitations for the locations where possibly infectious contacts take place, such as leisure activities, workplaces, and schools. As such, it is important to examine a model that allows for different strategies in locations with an agent model that models spatial networks.

There's a number of models that present an ABM to simulate infectious diseases such as COVID-19 and also include students and schools [11]. Many of these models, such as [7], only distinguish between open or closed schools without compromise solutions such as school rotation. In models such as [4], synthetic populations are used to examine different modes of school operations in combination with face-mask adherence. The number of students can be halved permanently, but students do not rotate weekly, which has different implications for the actual contact behaviors. In [13], a model is presented to examine the loss of schooling days due to school closures during the COVID-19 pandemic. Different strategies, such as reducing teacher-student ratios or the use of school rotation, are examined here. However, this model only considers households and a single school with multiple classes, which deviates from reality, where students attend different schools and interact across households, schools and classes during leisure activities.

Due to the desired flexibility in the range of questions that can be answered, we chose to use the SoSAD modeling approach, which allows for the simulation of different non-pharmaceutical interventions, spatial networks that support location-based policies and the inclusion of real-world data.

3   The SoSAD Modeling Approach

The SoSAD modeling approach aims towards flexibility and extensibility to allow swift response to new demands and developments. In the following sections, an overview of the key concepts will be given, starting with the modeling of the population and infrastructure, the activities and contacts during which contagion can take place and the countermeasures supported by the model. The conceptual behavior of the agents is described, while the implementation of the mechanisms around routines, interactions and contagions is displayed in Fig. 1 in a simplified manner focused on the activities of agents.

3.1   Population and Infrastructure

The agents are modeled after the general population of a German city with approximately 100.000 inhabitants. Thanks to the close cooperation with the city's local government, the researchers have access to anonymized data that provides information about the structure of households, schools and city districts. Each agent in SoSAD represents an individual person of a particular age. Depending on their age, these agents are clustered into three distinct behavioral groups: children (including adolescents), workers (including university students), and pensioners (i.e., all agents above the age of retirement). The population consists of a total of 102798 agents, of which 15888 are students, 67169 are workers and 19741 are pensioners. The infrastructure of the model consists of several locations, such as households, leisure activities, workplaces and schools as well as hospitals with attached intensive care units to include the pandemic's impact on the healthcare system. In total, 56663 households, 175 leisure activities, 175 workplaces and 33 schools with 400 classes are represented within the model. The number of households, schools and classes is based on official data.

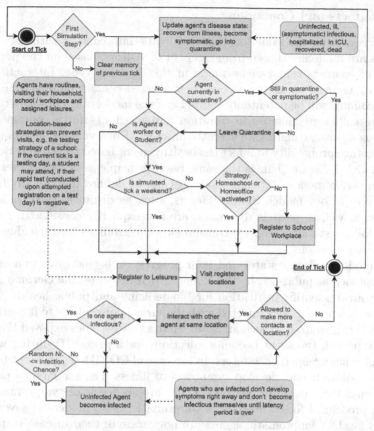

Fig. 1. The SoSAD modeling approach: decision-making of agents for daily activities. The rounded squares provide additional information about further mechanisms.

Agents and locations form a dynamic bipartite graph. This graph determines which agents can encounter and possibly infect each other at which location. The locations an agent frequents are determined by its daily routines defined during initialisation. Workers have a workplace which can represent a private company, a public service agency, as well as a university or other facilities. All agents under the age of 18 attend schools. Agents of any age have a household where they live alone or with other agents, depending on the population data. Furthermore, all agent's frequent leisure facilities which represent shops as suppliers of both essential and non-essential goods as well as cinemas, gyms, stadiums, and concert halls as well as any other public places for recreational activities. While not considered in this paper, special locations and infrastructure, such as school busses and public swimming pools, have been implemented and analyzed by request of the city to investigate the possible impact of policy decisions.

3.2   Contacts and Contagion

Agents frequenting the same location may come into contact with each other. Contacts are randomly chosen from the pool of visitors, based on the permitted number of contacts that can be made in this location type. Interactions are reciprocal, assuring no agents exceed their allowed contact numbers. Since not every encounter will be potentially contagious, the model only considers contacts that were sufficiently intense in duration and proximity for a contagion. If a contagious agent encounters a currently uninfected but vulnerable agent, there is a percentage probability to infect the healthy agent based on the infectiousness of the modeled disease. This is the same basic principle as in most other agent-based contagion models [11]. These settings can be defined individually for each agent group in our model. Spatial factors, such as distance, time, indoors or outdoors, as well as particle dynamics, are not explicitly considered. Infection chances follow estimated average transmission probabilities for typical activities at particular locations.

The model of disease states and their progression is analogous to a modified SEIR approach as published by the Robert Koch-Institute, the German government's central scientific institution for biomedicine and public health [2]. Any agent that has not yet been infected with the virus is susceptible to it (state S). If the virus is transmitted to such an agent, that agent becomes exposed (E). After a latency period, the agent becomes infectious for a period (I) during which it can infect other agents it encounters. In the case of COVID-19, an agent becomes infectious before it may develop symptoms of illness (i.e., the latency period is shorter than the incubation period). There are six levels of symptoms, one of which is predefined for each agent: asymptomatic, mild, moderate, severe, critical, and fatal. Asymptomatic agents are not aware of their disease state. Mild infections are not necessarily recognized as an infection with COVID-19 and an agent may continue going about their schedules despite minor symptoms [5]. Moderate symptoms mean that the agent may or must stay at home until it recovers, thereby having no further contacts with other agents at work, school, or leisure facilities. However, these agents will still interact with any other agent living in their household. Agents with severe or critical symptoms will be hospitalised, possibly with intensive care, and will not have any contacts during their stay. Agents with a fatal level of severity pass away and are removed from the model. Recovered agents will become (partially) immunized to further infection (R). Due to recent findings in the pandemic [1,9,15], recovery will decrease the reinfection chance of partially immunized agents and further assure that if a recovered agent is reinfected, their disease will be of decreased severity.

3.3   Activities and Countermeasures

Without any countermeasures to combat the spread of the virus, the disease will keep spreading repeatedly, although hypothetically, after a sufficient number of infections, any agent should either pass away or become fully immune. However, in reality, agents and (local) government and businesses will impose restrictions on the behavior of agents to slow down and reduce the infection dynamics.

The SoSAD model offers the following strategies and measures to influence the rate at which infections spread in the model: by forcing symptomatically ill patients into quarantine, the spread of the virus can be restricted to household members only, where infection is not necessarily guaranteed due to different living circumstances. By reducing the leisure contacts for both adults and children, infections can be reduced. This includes customer limits in stores, mandatory hygiene concepts at leisure facilities and reduced contacts with friends or family.

Once vaccines became widely available, Germany implemented the so-called '3G-Strategy': vaccinated, recovered, or tested (Geimpft, Genesen, Getestet). Only individuals with a valid vaccination, proven recovery or recent test result may access leisure activities such as restaurants, sport events and similar. Towards Winter 2021, the strategy was narrowed down to the two variants '2G', which no longer accepted unvaccinated and unrecovered individuals regardless of test results, as well as '2G Plus' which required a recent test result on top of vaccination or recovery certificate. These strategies are also present in the SoSAD model, allowing to account for the effects of such strategies on the infection dynamics at leisure activities and workplaces. While not all industries allow for the same degree of remote work, increasing the home office rate among the working population also helps reducing contacts in the workplace. In Fall 2021, Germany saw an estimated home office rate of about 20% [8] due to the accelerating infection dynamics.

In the same vein, homeschooling is another means of reducing contacts among children, either by fully closing schools or by having a certain percentage of children being homeschooled. School Rotation is a special form of schooling, in which classes are split in half and have students taking turns between in-person classes and online lessons. Another means of reducing the disease spread in schools is regular testing of students using rapid tests and the quarantine of students who were tested positive, along with classmates who have frequent contacts with them. Finally, schools with offset start times help reducing possibly infectious contacts among students on their way to school, given that public transportation may frequently be crowded. Social distancing cannot be guaranteed in such cases.

4   Simulating Three Schooling Strategies

In our analysis of schooling strategies, the following three options were simulated and evaluated:

(i) *Regular schooling with reduced contacts:* In this case, the regular class and course cohorts in the schools are taught completely as in normal operation. Distance rules can only be observed to a limited extent during lessons (depending on the room capacity). Therefore, mouth and nose protection are also worn during lessons and the room is aired regularly. The cohorts remain separated as much as possible during break times. However, complete separation is also not possible because of school bus traffic, so that infections may also occur across cohorts.

(ii) *Closed Schools:* In this case, there is no attendance at the schools. The schools are therefore eliminated as a site of possible infections.

(iii) *Rotation of halved classes (school rotation):* In this case, the class or course cohorts are divided in half. One half is taught in face-to-face classes and the other half is taught at home. The change takes place weekly. All other measures according to strategy (i) remain in effect here as well. Since less students meet on their way to school and less space is taken in the classes and other areas in school, contacts and infections among students are expected to be lower compared to schools operating at their usual capacity.

While different scenarios regarding the virus variant, contact rates and other circumstances have been simulated, this paper presents the results of a simulation study conducted using a highly infectious variant of COVID-19 inspired by the novel Omicron strain which causes skyrocketing infection cases in many countries. To model the high infectiousness of the Omicron variant [6], the model assumes the virus to be twice as infectious as the Delta variant and an increased reinfection rate of 50%, meaning that initially vaccinated people are no longer considered to be fully immune. In December 2021, researchers were not yet certain about the effectiveness of vaccinations against the new strain [6], inspiring the choice to set initial vaccinations to 0% to examine the impact of schooling strategy decisions in a worst-case scenario. The other parameters were calibrated using simulated annealing. Several configurations were able to replicate real world data. However, some combinations, such as very high leisure contacts for adults, contradict existing research [12] and thus, the authors chose a configuration that is consistent with empirical findings. In all three scenarios, the initial state is based on the month of December 2021 in Germany, based on official data provided by the RKI during that time period [14].

Due to the relatively low infection numbers in the model city over the course of the pandemic, the infections prior to the start of the simulation are based on reports of the corresponding federal state adjusted for the smaller population size. Due to the pandemic, reduced contact rates of agents are assumed compared to a non-pandemic [12]. Both private and professional contacts of adults are set to an average of two contacts per day. For students, a higher number of leisure and school contacts is assumed (number of contacts: 3 per day) [12]. This is partially due to the fact that in school buildings, space is often too limited to allow for effective social distancing. To ensure the safety of students, frequent tests are conducted to filter out infected students as early as possible. In this experiment setup, students attending school are tested twice in a 5-day-school week (on Monday and Thursday) using a rapid test. In case of a positive result, either due to infection or a false positive, the student is quarantined. This simplified testing strategy will be employed for any student attending school on testing days in both the open school and school rotation scenarios. The home office rate for workers and university students is rounded to an estimated 20% of the working age population working from home.

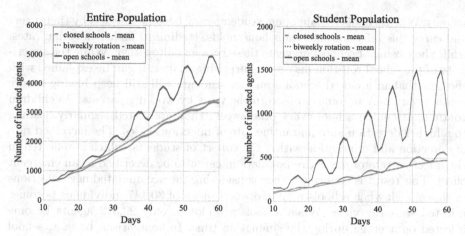

Fig. 2. Absolute number of active infections among the entire population (left) and students (right) with confidence interval (95%). Measured average of 100 runs.

The only variation between the three scenarios lies in the schooling strategies. For the school rotation scenario, the school contacts are additionally reduced to 2 to express the fact that fewer students attend school. This leads to easier social distancing in classrooms, fewer individuals using public transportation and splitting of social groups. In both open schools and school rotation, testing is applied as described above. The three scenarios were run with 100 different random seeds over 60 ticks each, only varying the parameters regarding the schooling strategies. The ticks represent one day within the simulation. The first ten ticks are considered a warm-up period in which the model initializes based on input data regarding infections among different cohorts at start. Thus, these first ten ticks were discarded and removed from the evaluation.

5 Can We Keep Schools Open? Simulation Results

To examine the impact of the three proposed schooling strategies, the number of infected agents was plotted both on the level of the general population as well as on student-level only. Figure 2 depicts the impact of the schooling strategies on the entire population (left) and students (right): Students may become sick in all scenarios, but both school rotation and the closing of schools are effective at reducing infections.

Depending on the strategy, the graphs represent a wave-shaped infection curve. The local minima represent weekends when there are fewer infectious encounters between agents due to the lack of professional contacts (workplace contacts and school contacts).

A surprising result of this study is that school rotation appears to be even more effective at lowering infections among the general population than closed schools. When schools remain open, students may become infected by their peers and carry this infection into their households, leading to higher infection rates, while the closing of schools prevents these contacts altogether. While the superiority of the school rotation may seem surprising at first, it can be explained with the fact that in a closed school scenario, students will still keep having leisure contacts and may become exposed to the virus by working parents. As children continue attending school every other week, they are tested regularly, leading to a higher detection and quarantine rate of infectious cases. The increased rate of detection and quarantine within the cohort of students, which is reflected in the infection rate of the entire population, can also be described as an emergent effect. This result is confirmed when considering the accumulated new infections in this model: when schools remain open, a mean of 30,147 individuals becomes infected with the virus. When schools are closed, only 23,512 agents become infected on average during the simulation time. In comparison to that, school rotation proves narrowly superior with only 22,401 infections on average during the simulation. Thus, the numbers confirm the conclusion drawn from the visualisations. While the difference between closed schools and school rotation strategies is small in terms of total infection numbers, it is important to remember that studies have proven the negative effects of closed schools [10], meaning that school rotation may provide a compromise solution.

The choice of schooling strategies has a strong impact on both the overall population and the student population. As the results show, reducing school operations alone is not sufficient to contain the pandemic. Still, students in particular benefit from the change in school strategy from open schools to school rotation without having to close the school entirely. Switching to school rotation also has the advantage that students are additionally tested when they attend school. Infected students can therefore be detected and quarantined, preventing further infections during leisure activities. When considering the entire population, alternating operation is even superior to school closure. This result showcases the special characteristic of ABMs: the ability to discover patterns that emerge from a combination of mechanisms without explicit modeling.

6 Discussion – Patterns in Different Experiment Setups

While only the results of one simulation experiment setup were discussed, more experiments were conducted in the past months when advising various crisis response groups. The most important takeaway of these simulation studies is the pattern shown above: school rotation shows a similar effectiveness in reducing infections as closing schools, as well as flattening the wavy behavior of the open school infection graphs.

In December 2021, during the first observation of the novel Omicron variant, different worst-case scenarios regarding the infectiousness and immune escape potential of the Omicron variant were examined. Even when the traits of the virus were greatly exaggerated compared to the observations in the real world, the general pattern of infection-mitigating effects of school rotation held up.

Further experiments in the middle of December 2021, examining the effects of various lockdown scenarios over Christmas and New Year's, confirmed the same result patterns. In the lockdown scenario, in addition to switching the school strategies, other measures were considered, such as company vacations, various home office rates as well as contact restrictions. Other experimental setups in which the effect could be reproduced include combining the strategies with different home office rates, different vaccination rates, and with different COVID-19 variants, such as the original variant, Delta and possible variations of Omicron based on first published and rumored estimations. The positive effect of the school rotation on the entire population as well as on the students themselves can be reproduced.

As mentioned above, several parameter configurations were found that replicate infection patterns matching real world observations, including some that match further empirical findings. Therefore the model is generally plausible in its ability to produce realistic behavior. Overall, the model still needs further testing and validation. Calibration has shown that the model is generally capable of producing realistic behavior, which lends some level of credence to the results and trends shown up in the simulation studies. Given the setup of the experiments in which other parameters, such as contact rates, home office strategies among adults and even disease traits, have shown consistent patterns, school rotation appears to have positive effects on the population and students both. These effects are robust to parameter changes, though it is still up to decision makers to determine whether the difference between closed and school rotation operations is acceptable in the given situation.

Systematic real-life experiments between schooling strategies would be the best means of validating the model, but such experiments are not practically feasible due to ethical reasons. Further, since governments typically present several measures at the same time, it is difficult to separate the effects of different combinations into the contributions of individual policy decisions to compare the three different strategies. In such situations, sufficiently plausible and realistic simulations can help distinguishing the effects of strategies and attributing observations to individual measures.

7   Conclusion – What Only ABMs Can Show

This paper presented an agent-based model to simulate the spread of a disease in a population. In this case, the model refers to the COVID-19 virus, which is spread when agents interact in different locations such as households, workplaces, schools and leisure activities. This model is used to simulate and analyze different schooling strategies to slow down and reduce infections among students. The experiments compared the impact of open schools with closed schools and school rotation, in which classes are halved and take weekly turns between in-person attendance and online schooling. The experiments have shown that school rotation is not only superior to open schools in terms of preventing infections, but even comparable to closed schools and may outperform the closed school strategy on a population level due to continued monitoring and quarantining of infected students through regular testing.

Schools are often said to not have a major impact on the infection dynamics across the entire population, but it is still important to prevent harm from children, a vulnerable demographic. Given the need to balance different interests, ABMs can help making such decisions – while both working parents and children would certainly favor open schools, it may be possible to reach a point in which keeping the schools open is considered an irresponsible decision. As such, the school rotation strategy prevents schools shutting down completely. A statistical model might have predicted that school rotation strategies offer a compromise between leaving schools open or closing them, but ABMs are superior in their ability to express the impact of an intervention on specific population groups. The key difference between statistical models and ABMs is the possibility to model individual activities, household structures and dynamic contact graphs. Infections can spread non-uniformly, leading to emergent behavior showcased in the results of this paper.

Without an ABM, the effects of school rotation would likely be dismissed entirely, given that the benefits and drawbacks of some approaches are difficult to conceptualize. Emergent behaviors such as this are often difficult to anticipate. The positive effect of school rotation on detection and quarantine further emphasizes the value of such complex models, given that a simpler model without different locations, agent groups and strategies would not have the capacity to show such emergent effects. Therefore, the authors believe that ABMs are a valuable tool in policy-making not just in the pandemic, but in any situation in which some decisions may show only little effects on a large scale but important impact on specific population groups which may be overlooked otherwise. In the future, the model will be extended by further components and also further validated. For this purpose, additional cities will be integrated into the model to test the model behavior in relation to other structural and demographic circumstances and the transferability to cities with different population and infrastructure.

Acknowledgement. This model was created in the context of *AScore*, a consortium project funded from 01/2021 until 12/2021 within the special program "Zivile Sicherheit - Forschungsansätze zur Aufarbeitung der Corona-Pandemie" by the German Federal Ministry of Education and Research (BMBF) under grant number 13N15663.

References

1. Abu-Raddad, L.J., et al.: Severity of SARS-CoV-2 reinfections as compared with primary infections. N. Engl. J. Med. **385**(26), 2487–2489 (2021). https://doi.org/10.1056/NEJMc2108120
2. Buchholz, U., et al.: Modellierung von Beispielszenarien der SARS-CoV-2-Ausbreitung und Schwere in Deutschland (2020)
3. Chowdhury, M.J.M., et al.: COVID-19 contact tracing: challenges and future directions. IEEE Access **8**, 225703–225729 (2020). https://doi.org/10.1109/ACCESS.2020.3036718

4. España, G., et al.: Impacts of K-12 school reopening on the COVID-19 epidemic in Indiana, USA (2020). https://doi.org/10.1101/2020.08.22.20179960
5. Espinoza, B., et al.: Asymptomatic individuals can increase the final epidemic size under adaptive human behavior. Sci. Rep. **11**(1), 1–12 (2021). https://doi.org/10.1038/s41598-021-98999-2
6. European Centre for Disease Prevention and Control: Assessment of the further emergence and potential impact of the SARS-CoV-2 Omicron variant of concern in the context of ongoing transmission of the Delta variant of concern in the EU/EEA, 18th update (2021). https://www.ecdc.europa.eu/en/publications-data/covid-19-assessment-further-emergence-omicron-18th-risk-assessment
7. Ghorbani, A., et al.: The ASSOCC simulation model: a response to the community call for the COVID-19 pandemic. Rev. Artif. Soc. Soc. Simul. (2020). https://rofasss.org/2020/04/25/the-assocc-simulation-model/
8. Google: COVID-19 Community Mobility Reports. https://www.google.com/covid19/mobility/index.html. Accessed 28 Jan 2022
9. Hall, V.J., et al.: SARS-CoV-2 infection rates of antibody-positive compared with antibody-negative health-care workers in England: a large, multicentre, prospective cohort study (SIREN). Lancet **397**(10283), 1459–1469 (2021). https://doi.org/10.1016/S0140-6736(21)00675-9
10. Lee, J.: Mental health effects of school closures during COVID-19. Lancet Child Adolesc. Health **4**(6), 421 (2020). https://doi.org/10.1016/S2352-4642(20)30109-7
11. Lorig, F., et al.: Agent-based social simulation of the COVID-19 pandemic: a systematic review. J. Artif. Soc. Soc. Simul. **24**(3), 5 (2021). https://doi.org/10.18564/jasss.4601
12. Mossong, J., et al.: Social contacts and mixing patterns relevant to the spread of infectious diseases. PLoS Med. **5**(3), 0381–0391 (2008). https://doi.org/10.1371/journal.pmed.0050074
13. Phillips, B., et al.: Model-based projections for COVID-19 outbreak size and student-days lost to closure in Ontario childcare centers and primary schools (2020). https://doi.org/10.1101/2020.08.07.20170407
14. Robert Koch-Institut: SARS-CoV-2 Infektionen in Deutschland, January 2022. https://doi.org/10.5281/zenodo.5908707
15. Schuler, C.F., IV., et al.: Mild SARS-CoV-2 illness is not associated with reinfections and provides persistent spike, nucleocapsid, and virus-neutralizing antibodies. Microbiol. Spectrum **9**(2), e00087-21 (2021). https://doi.org/10.1128/Spectrum.00087-21
16. Shinde, G.R., Kalamkar, A.B., Mahalle, P.N., Dey, N., Chaki, J., Hassanien, A.E.: Forecasting models for coronavirus disease (COVID-19): a survey of the state-of-the-art. SN Comput. Sci. **1**(4), 1–15 (2020). https://doi.org/10.1007/s42979-020-00209-9
17. Squazzoni, F., et al.: Computational models that matter during a global pandemic outbreak: a call to action. J. Artif. Soc. Soc. Simul. **23**(2), 10 (2020). https://doi.org/10.18564/jasss.4298
18. Timm, I.J., et al.: Kognitive Sozialsimulation für das COVID-19-Krisenmanagement - Social Simulation for Analysis of Infectious Disease Control (SoSAD). Technical report, Deutsches Forschungszentrum für Künstliche Intelligenz (DFKI), August 2020

Data-Driven Agent-Based Model Development to Support Human-Centric Transit-Oriented Design

Liu Yang[1,2(✉)] [iD] and Koen H. van Dam[3] [iD]

[1] School of Architecture, Southeast University, Nanjing 210096, China
yangliu2020@seu.edu.cn
[2] Research Centre of Urban Design, Southeast University, Nanjing, China
[3] Centre for Process Systems Engineering, Department of Chemical Engineering, Imperial College London, London, UK
k.van-dam@imperial.ac.uk

Abstract. This paper proposes an agent-based simulation model of activities within an urban environment to evaluate alternative transport-oriented development (TOD) designs and infrastructure investment proposals prepared by urban planners. The students test the model as model users, and the generated model output on the use of the city infrastructure, occupancy of public space, and key data around the pedestrian and vehicle movements can be translated to design modifications by comparing results with desired targets. This provides valuable scenarios to key stakeholders in the design. A particular challenge with using simulation models as part of the decision-making process is the need to include realistic data for the behaviour of the transport system users. To this end, an experiment was conducted in which data on the individual behaviour and activities was collected, which can be integrated into the simulation model to capture realistic responses to TOD proposals. Illustrative results are shown, demonstrating the model can produce meaningful results for planners but also highlights the role of agent-based simulation models in steering the data collection process and engaging with decision-makers.

Keywords: TOD · Agent-based model · Human factors · Data collection · Decision-support tool

1 Introduction

Integrated design of transit stations and affiliated urban areas such as transit-oriented development (TOD) has gained increasing attention worldwide. TOD is regarded as a primary principle of urban planning [1], especially to minimize dependencies on cars and promote more sustainable modes of transport as part of the transition in cities to

---

The original version of this chapter was revised: The acknowledgement section has been updated. A correction to this chapter can be found at https://doi.org/10.1007/978-3-031-20179-0_10

© Springer Nature Switzerland AG 2022, corrected publication 2024
F. S. Melo and F. Fang (Eds.): AAMAS 2022 Workshops, LNAI 13441, pp. 60–66, 2022.
https://doi.org/10.1007/978-3-031-20179-0_3

lower emissions and improve local conditions. The design of a new generation TOD thus emphasizes improving access to active travel and high-quality public space to promote human comfort [2]. Therefore, there is a need for urban design support tools with a fine spatial-temporal resolution which can be used to examine how people use the infrastructure and public space, so designers can evaluate different plans against key metrics and discuss options with relevant stakeholders. Agent-based modelling (ABM) is a suitable methodology to create a heterogeneous population with activity patterns that lead to transport decisions (including mode, route, and time) for a given environment and infrastructure options. Collectively, the individual decisions lead to insights on key indicators that can support planners in evaluating different alternatives and ensure new developments are attractive and efficient for users but also meet sustainability and economic targets. One of the challenges for ABM applications in this domain is building data-driven models and exploring human behaviour, which can be integrated within the model to study responses to various interventions. This paper explores this theme for a case study in Nanjing, China.

Model development itself can help guide data collection [3] by showing what data is required to test a theory. Moreover, models are often built as generic frameworks, which are then instantiated for a specific case study by providing relevant case-specific input data. Such data helps to test a few implicit and explicit assumptions about behaviour [4]. The ODD protocol ("Overview, Design concepts, Details") designed to describe ABMs refers explicitly to this as "initialization" and "input data" to describe part of the model "if scenarios with different initial conditions are implemented or if input data are essential drivers of model dynamics" [5].

However, there remains several methodological challenges, for example, in collecting data that matches the specification of the model, linking data sets together, analyzing the data to extract significant drivers and behaviours [6], deriving agent rules from data, and integrating human-environment models [7]. In addition, Kagho et al. highlighted that "the data collection process is one-way error can be introduced into the model", and data bias (e.g., introduced by preference surveys) could cause bias in models [8].

This paper, therefore, aims to: 1) build a data-driven ABM decision-support tool for urban designers, especially in designing and evaluating people-centric TOD plans; and 2) discuss the role of data in the development of the urban simulation model, and the use of output data to help influence decision-making in a case study in Nanjing, China.

2 A Prototype ABM

To meet these aims, we firstly developed a prototype model, "Transport, Spaces, and Humans-system (TSH-system)", and implemented the model in the GAMA platform (documented in https://gama-platform.org/wiki/Projects). Figure 1 shows the model interface, providing options for the users to load datasets and TOD designs and run the simulation to visualize the movement of the agents within this space, collecting relevant data. The model was built as a generic framework to support students and practitioners in urban planning and design, architectural design, and other fields to analyse urban systems and quantitatively evaluate design schemes [9, 10]. It allows the simulation of private car drivers and pedestrians for a given TOD plan to predict the usage ·

of the space and relevant activities, as well as automobile travel demands, active travel demands, and transport mode choices. The designers can then interpret these insights by making changes where necessary and checking the effect of those in new model runs.

Input data, provided by the model users, includes GIS files (land use, walking routes, and driving road network), population statistics (e.g., density), activity patterns, walking and driving speeds, mode choice parameters (e.g., the weight of money cost in mode choice), personal parameters (e.g., shoulder width), and pedestrian parameters (e.g., the repulsive force in social force model). The model then outputs hourly data regarding users over the urban space (occupancy/dwelling time), automobile traffic volumes and pedestrian population on each road segment.

Fig. 1. Interface of the TSH-system model

3   Data Collection

An experiment was set up in Nanjing, China to collect relevant behavioural data. The experiment aims to explore the pedestrians' walking behaviour in surrounding areas of rail transit stations as well as the impact of the design of such areas (underground/semi-underground/open outdoor space) on their behaviour, cognition, and comfort. Thirty-four participants were recruited, and four of them only attended a pre-test experiment. Each participant visited the different spaces of three subway stations and one railway station freely for ten minutes while being monitored remotely by the researchers.

The factors of time-space trajectories, electroencephalography (EEG), eye movement, electrodermal activity (EDA), and heart rate variability (HRV) were investigated. The ErgoLAB human-machine environment platform (Kingfar International Inc. Beijing, China) and a series of wearable physiological recording modules were used to synchronously collect and analyse multi-dimensional human factors data. Furthermore, self-reported subjective assessments were collected by a survey. In our sister paper [11], we analysed the initial results of EEG and eye tracking. We found that the human brain becomes more active with attention more scattered when walking from the semi-underground space to ground level.

For the first step of building a data-driven ABM, we extracted features of pedestrians' walking behaviour from the time-space trajectories, including the movement direction angle. This would inform the behaviour of the agents, making the model more realistic compared to a previous implementation without such data but based on more general high-level characteristics, and also served as a first test to incorporate this kind of experimental data in the agent rules and parameters.

4 Case Study

The TSH-system model was tested in a regular MSc course at the Southeast University in China, which attempts to explore the future development model of the surrounding plots of transit stations [8]. This year, a case study was conducted in Shanghai, China. The research site is located at plots X101-01 and X102-02 of Shanghai West Railway Station (see Fig. 2). It is not only the location of the Shanghai West Railway Station but also a transfer station for Metro Lines 11, 15 and 20; thus, it is an essential hub for the local and wider area.

The ABM was initiated with shared data: The GIS data was based on a baseline of the built environment, with the modifications and designs prepared by the user (i.e., the urban design students) as part of their proposed intervention. In addition to the spatial data, agent-behaviour data obtained from the experiment was used to simulate the users of the urban system.

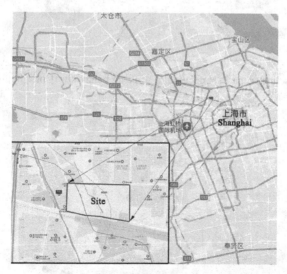

Fig. 2. Case study site in Shanghai, China

Using the TSH-system agent-based model for this case study, Fig. 3 shows the initial results of simulating the users of urban spaces (blue plots) and walking behaviour (green lines) for the baseline scenario, simulating how people in the city interact with the current TOD layout.

On this basis, students conceived their designs following a primary aim, for instance, to improve spatial orientation and wayfinding, to combine two grid road systems, and to create a high-quality microclimate. Each student prepared, documented and justified their TOD plans against these aims. To test the effectiveness of their plans, they changed the GIS input files in terms of the land uses, activities, road network and pavement network, and ran the simulation model for their specific scenario. The model then presented the hourly number of people in urban spaces, traffic volumes over the road network, and walking demands across the pavement network, which provided the designers with relevant metrics to help them revise their plans as well as their narrative for the chosen approach iteratively (see Fig. 4). This simulation model could easily be used to test multiple designs, if desired, and show how non-modellers can benefit from an agent-based simulation model implemented by others, without having to modify code to experiment with the system.

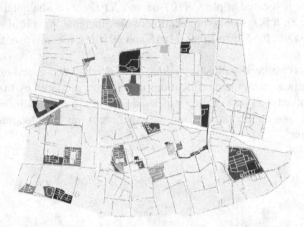

Fig. 3. Baseline scenario simulation of walking behaviour (Color figure online)

Fig. 4. A TOD design scenario simulation of walking behaviour

5 Discussions and Conclusion

These initial results illustrate the potential of enriching the prototype TSH-system model with the data from the experiment to generate a more reliable output for evaluating a given design. This enables designers to compare alternatives for the physical design of TOD projects for a given population and environment. Besides, the time-space trajectories and physiological and psychological data collected in the experiments match the model's specification, that is, simulating human behaviour in public spaces around transit stations. Furthermore, even though the investigation took place in a different city, the changes to the model implementation meant it could be applied elsewhere under the assumption that people use given infrastructure similarly. Given that the two cities are neighbouring cities in the same region this seems a fair assumption, though, for example in an international application, this might need to be verified. Also, the way of the collection was designed to avoid data bias by not only delivering surveys but also recording individual behavioural data using wearable physiological recording devices. This means that while the participants were aware that they were in an experiment, there were multiple ways to match data and check responses.

To incorporate the collected data into the model, we are analysing the data to extract significant drivers of individual behaviour in the TSH system and derive agent rules and behaviour-related parameters. As always with such complex systems, there is uncertainty around key input parameters, especially when these are based on an analysis of human behaviour. In the next stage of this project, we can use sensitivity analysis to test these parameters' impact on the final result and use that to guide the design of updated data collection strategies and experimental setup. For TOD, this specifically refers to mode choice and journey purpose, but also the agent's views on the quality and attractiveness of the space. The role of culture in these views could also be incorporated into further work by running similar experiments in different locations (e.g., varying city size, country, and climate).

There are, however, some challenges in developing data-driven ABMs. For example, preparing and cleaning the GIS files before integrating with agent-based models is time-consuming. Standardization of data formats, quality checks, and scaling up data to the population level are also challenging issues. To this end, we aim to integrate this work with a geospatial data platform to take advantage of other relevant datasets (e.g., on the environment) and to present simulated data on the platform. This way, datasets that have been prepared and tested previously can be taken into account, also improving consistency with results based on other analyses of the same input data. Making this more interoperable would also enable a more efficient comparison of results using data from different years or providers, for example.

In the next stage of this project, additional alternative designs will be tested by the students to extract insights from urban designers as model users and understand how they translate the generated insights into design modifications as well as how they use the model as part of the rationale for a proposed project. The model users will be asked to complete a survey on their experience using the model, comparing this with some challenges stated before the engagement, to quantify the benefits of having such a data-driven model available as part of the design trajectory. This way, the model can be used in discussions with relevant stakeholders, including key decision-makers, but also with the users of the space in engagement activities, and finally, to further validate the data by combining experiments, simulation results, and real-world experiences of experts. As such, the approach presented here can help designers evaluate their work, modellers improve the behavioural aspects of the agent-based urban simulation, and drive data-collection strategies for experimental design.

Acknowledgements. Liu Yang is supported by the National Natural Science Foundation China (No. 52108046), Natural Science Foundation of Jiangsu Province (No. BK20210260), and China Postdoctoral Science Foundation (No. 2021M690612). Koen van Dam works on the UK FCDO funded programme Climate Compatible Growth. The project was partly funded by the University-Industry Collaborative Education Program of the Ministry of Education (No. 202101042020). The authors thank Dr Yuan Zhu (Southeast University, China) for leading the MSc course and supporting the experiment. We thank Ms Yanmeng Wang for preparing the site figure. Finally, Liu would like to thank Dr Patrick Taillandier, Dr Benoit Gaudou, and Mr Tu Dang Huu for their assistance in model development.

References

1. Calthorpe, P.: The Next American Metropolis: Ecology, Community, and the American Dream. Princeton Architectural Press, New York (1993)
2. Cervero, R., Guerra, E., Al, S.: Beyond Mobility: Planning Cities for People and Places. Island Press, Washington, DC (2017)
3. Epstein, J.M.: Why model? J. Artif. Soc. Soc. Simul. **11**(4), 12 (2008)
4. Crooks, A., Castle, C., Batty, M.: Key challenges in agent-based modelling for geospatial simulation. Comput. Environ. Urban Syst. **32**(6), 417–430 (2008)
5. Yang, L., Bustos-Turu, G., van Dam, K.H.: Re-implementing an agent-based model of urban systems in GAMA. In: GAMA Days 2021, 23–25 June 2021, p. 43 (2021)
6. Heppenstall, A., Crooks, A., Malleson, N., Manley, E., Ge, J., Batty, M.: Future developments in geographical agent-based models: challenges and opportunities. Geogr. Anal. **53**(1), 76–91 (2021)
7. An, L., et al.: Challenges, tasks, and opportunities in modeling agent-based complex systems. Ecol. Model. **457**, 109685 (2021)
8. Kagho, G.O., Balac, M., Axhausen, K.W.: Agent-based models in transport planning: current state, issues, and expectations. Procedia Comput. Sci. **170**, 726–732 (2020)
9. Yang, L., Zhu, Y, van Dam, K.H.: Supporting the use of agent-based simulation models by non-modeller urban planners and architects. In: Social Simulation Conference, 20–24 September 2021 (2021)
10. Grimm, V., et al.: The ODD protocol for describing agent-based and other simulation models: a second update to improve clarity, replication, and structural realism. J. Artif. Soc. Soc. Simul. **23**(2) (2020)
11. Yang, L., Zhu, Y., Chatzimichailidou, M.: Ergonomics analysis of the pedestrian environment around subway stations. In: CEB-ASC2022, The 1st Environment and Behavior International Symposium, 20 November 2022 (2022)

Enabling Negotiating Agents to Explore Very Large Outcome Spaces

Thimjo Koça[1]([⊠]) [iD], Catholijn M. Jonker[2,3] [iD], and Tim Baarslag[1,4] [iD]

[1] Centrum Wiskunde & Informatica, Amsterdam, The Netherlands
{thimjo.koca,T.Baarslag}@cwi.nl
[2] TU Delft, Delft, The Netherlands
c.m.jonker@tudelft.nl
[3] Leiden University, Leiden, The Netherlands
[4] Utrecht University, Utrecht, The Netherlands

Abstract. This work presents BIDS (**Bi**dding using **D**iversified **S**earch), an algorithm that can be used by negotiating agents to search very large outcome spaces. BIDS provides a balance between being rapid, accurate, diverse, and scalable search, allowing agents to search spaces with as many as 10^{250} possible outcomes on very run-of-the-mill hardware. We show that our algorithm can be used to respond to the three most common search queries employed by 87% of all agents from the Automated Negotiating Agents Competition. Furthermore, we validate one of our techniques by integrating it into negotiation platform GeniusWeb, to enable existing state-of-the-art agents (and future agents) to scale their use to very large outcome spaces.

Keywords: Automated negotiation · Very large negotiation domain · Search

1 Introduction

Over the last decades, more and more processes and information are being digitized, allowing the implementation of new technologies that can reduce the duration and complexity of business processes. Automated negotiations is a promising example of such technologies that can bring benefits to various fields, including procurement [8], supply chain management [24], and resource allocation [2].

In such fields, negotiations can take place over high number of finite issues (roughly 100 issues or more). For instance, suppose a buyer (Bob) is trying to negotiate with one of his suppliers (Sally) over the delivery of 100 shipments for the next year, as depicted in Fig. 1. For each shipment there are 365 possible delivery dates, resulting in an outcome space with 365^{100} possibilities. Every time Bob has to propose a new offer to Sally, he needs to define some criteria that his next offer must fulfill (e.g. bring him a certain level of utility, lie within a utility interval, conform some trade-off between his own preferences and Sally's) and then search the enormous outcome space for a bid that best fits his criteria.

© Springer Nature Switzerland AG 2022
F. S. Melo and F. Fang (Eds.): AAMAS 2022 Workshops, LNAI 13441, pp. 67–83, 2022.
https://doi.org/10.1007/978-3-031-20179-0_4

Moreover, since Bob and Sally keep their preferences private, to increase the chances in achieving an agreement, he needs to: (a) exchange a high number of offers with Sally; (b) propose offers that, over time, are qualitatively as diverse as possible, i.e. sample broadly the outcome space. Hence, Bob needs a *scalable* way to search the enormous outcome space in a manner that is *timely*, *accurate*, and *diverse*.

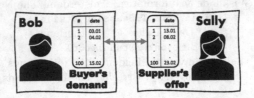

Fig. 1. A negotiation example.

However, searching a large discrete outcome space in the context of automated negotiation is a difficult task, mainly for two reasons. First, because the search process often translate to combinatorial problems that are impossible to solve exactly and challenging to solve approximately. Second, because it is not straightforward to design search algorithms that scale well, have accuracy guarantees, and explore the outcome space rapidly and in a diverse way, since there are trade-offs between the four properties.

Search mechanisms proposed by state-of-the-art agents and automated negotiation platforms perform poorly in very large outcome spaces (as we will see in Sect. 5), because of their underlying assumptions. In particular, they assume that: (a) either that the outcome space is small enough to be enumerated and explored rapidly [21,27], and are therefore not scalable; (b) or that the space can be randomly sampled [11,23,25], and as a result perform poorly in very large spaces; (c) or that search goals can be defined deterministically for each individual negotiation issue [18], which can lead to poor accuracy in finite domains as well as narrow exploration of the outcome space.

In this work, we propose BIDS (**B**idding using **D**iversified **S**earch) — an algorithm that can search outcome spaces with as many as 10^{250} possible outcomes given that the user's preferences are expressed by the widely-used additive utility function (i.e. with no issue interdependencies). The algorithm employs a dynamic-programming approach to exploit the additive structure of the utility function. We show that our methodology is accurate since it identifies approximate solutions with arbitrary error bounds to the search problem and can provide diversity since it is able to explore the outcome space broadly. Furthermore, we show that our methodology is generic by first surveying the search queries used by the agents that have participated in the Automated Negotiating Agents Competition (ANAC), and then use our algorithm to implement the three most common search queries employed by 87% of ANAC agents — the utility-lookup query, the utility-sampling query, and the trade-off query. Lastly, we validate

BIDS by integrating it into negotiation platform GeniusWeb so that state-of-the-art (and future) agents that need the utility-lookup query can use it.

2   Problem Setting

Our purpose in this work is to propose algorithms that tackle three search queries that are commonly used by negotiating agents in a manner that is scalable, rapid, accurate, and provides diversity. To do so, we need to first formalize each of the queries and discuss the associated challenges.

2.1   Negotiation Model

Agents in our setting negotiate over a finite set of issues $\mathcal{I} = \{1, \ldots, n\}$ and each issue $i \in \mathcal{I}$ has an associated finite set of values V_i. For instance, in the task-scheduling scenario of Fig. 1, the issues to negotiate upon are the 100 deliveries and the possible values per issue are the 365 d. All possible combinations of values form the *outcome space*, which is denoted by $\Omega = \prod_{i \in \mathcal{I}} V_i$. Each element $\omega \in \Omega$ is called a negotiation *outcome* and whenever it is convenient we will denote the component of ω corresponding to issue $i \in \mathcal{I}$ by $\omega_i \in V_i$.

The private preferences of each party over Ω are expressed through a utility function $u : \Omega \to [0, 1]$. We focus in this work on utility functions that are additive with respect to the utilities of each issue:

$$u(\omega) = \sum_{i \in \mathcal{I}} \lambda_i \cdot u_i(\omega_i)$$

where $\lambda_i \geq 0 \land \sum_{i \in \mathcal{I}} \lambda_i = 1$ are the weights defined for each issue, and $u_i : V_i \to [0, 1], \forall i \in \mathcal{I}$ are utility functions defined over each individual issue. There are no dependencies between individual issues since the utility function is a convex sum of individual issue utilities. The reasons why we picked additive utility functions are that they are widely used [4,6,14,15,28] and because, as we will see in Sect. 4, their structure allows for some scalable, rapid, accurate, and diverse search of the outcome space. Note that additive utility functions can code rather complex preference structures, since we do not make any assumptions on u_i, and as consequence we can define over each issue arbitrary utility functions (i.e. not necessarily linear).

A negotiation protocol (e.g. the Alternating Offers Protocol (AOP) [26]) regulates how the agents exchange offers during the negotiation. We consider protocols that allow in each round the communication of one or several bids, i.e. possible outcome(s) ω to agree upon, or a special message — for instance, a message that indicates acceptance of the opponent's latest offer, or a message informing a walk-away.

2.2 Typical Search Queries

There are many negotiating agents in literature, each with their own negotiation strategy and learning methodologies [5,11,12,18,19,24]. However, despite the richness of negotiating strategies, when an agent decides on a bid to propose next, it generally complies to the following pattern: It first sets some criteria that the proposed bid need to satisfy – examples include an appropriate utility target, a utility interval of interest, a trade-off between the own preferences and the opponent's — and subsequently tries to identify the most appropriate bids that meet one of three important search queries (illustrated in Fig. 2).

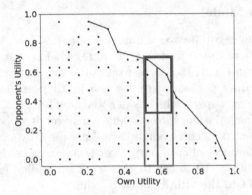

Fig. 2. Illustration of the three most common search queries over a utility diagram. The x-axis shows the agent's own utility and the y-axis the opponent's utility. Dots represent possible bids and the continuous curve represents the Pareto-frontier. While the blue line illustrates the possible picks for the utility-lookup query, the red rectangle illustrates the possible options for the utility-sampling query and the green rectangle the possibilities for the trade-off query. (Color figure online)

The utility-lookup is the simplest among the three queries. Agents define in each round a utility target $u_t \in \mathbb{R}$ and search for bid(s) with utility as close as possible to the target utility (illustrated by the blue line in Fig. 2). The target can be determined through a time-based strategy (e.g. Agent K [19]), a behavior-based strategy (e.g. Nice-Tic-For-Tac Agent [5]), or through some other criteria (e.g. through a resource-based tactic [13]). Formally, it can be defined as a minimization problem:

$$\underset{\omega \in \Omega}{\operatorname{argmin}} \quad |u(\omega) - u_t| \tag{1}$$

The second query is the sampling-utility query. In each round, agents search for one or more bids with utility that lies within a utility interval $[u_{min}, u_{max}] \subseteq \mathbb{R}$ (illustrated by the red rectangle in Fig. 2). Most works in literature determine the bounds of the interval through a time-based strategy (e.g. Agent M [25]). Formally, the query can be expressed as identifying a set S containing outcomes within a certain utility interval:

$$S \subseteq \{\omega \in \Omega : u(\omega) \in [u_{min}, u_{max}]\} \tag{2}$$

The third query considers some trade-offs while generating a new bid. Agents search for bids that optimize some objective function $f : \Omega \to \mathbb{R}$, while asking for at least a minimum utility for themselves (illustrated by the green rectangle in Fig. 2). Typical objective functions model the opponent's preferences in some way, for instance by estimating the opponent's utility function (e.g. The Fawkes Agent [20]) or by minimizing distance to the opponent's offers (e.g. Similarity-Tactic [12]). Formally, it can be defined as a constrained optimization problem:

$$\underset{\omega \in \Omega}{\text{argmin}} \quad f(\omega)$$
$$\text{subject to} \quad u(\omega) \geq u_t \tag{3}$$

The three queries are rather generic. In fact, when surveying the search queries used by the participants of the Automated Negotiating Agents Competition (ANAC) [6] since its inception (2010–2021), we find 87% of all participating agents use one of the three identified queries (see Table 1). Given their ubiquity, it is important to have a generic, well-founded way to answer these queries, either as part of a well-known negotiation framework (for instance [21], or [23]) or as a module available to future agents. This would aid to decouple the negotiation strategies of agents from their search methods, and as a result make the comparison of negotiation strategies easier.

Table 1. Typical search queries used by ANAC agents.

Search query	% of agents per ANAC year										total
	2021	2018	2017	2016	2015	2014	2013	2012	2011	2010	
Utility-lookup	23%	57%	14%	17%	4%	9%	11%	20%	33%	13%	20%
Utility-sampling	33%	19%	52%	78%	70%	50%	45%	40%	56%	29%	50%
Trade-off	33%	19%	5%	5%	13%	27%	22%	30%	11%	29%	17%
Other	11%	5%	29%	0%	13%	14%	22%	10%	0%	29%	13%

2.3 Design Specification of Search Algorithms Used by Negotiating Agents

Algorithms that can answer the three discussed queries need to provide a good balance between four specifications. First, the search need to be *scalable*, since we want agents to be able and negotiate in realistic scenarios, over even more than 100 issues (e.g. in fields such are supply-chain management and procurement). Second, algorithms need to be *accurate* so that negotiation strategies operate with minimal error. Third, since negotiation parties keep their preferences private, search algorithms need to be *rapid* so that the agents can exchange a high number of offers and therefore increase their chances in achieving an agreement. Four, the algorithms need to provide *diversity*, so that there are higher chances to achieve an agreement that is Pareto efficient (i.e. a win-win agreement) even though the opponent's preferences are not known.

Example 1 (Importance of diversity). To understand the importance of diversity suppose a buyer negotiates with the seller to obtain a TV considering price {low, average, high} and quality {low, moderate, high}. The buyer aspires for a high-quality TV at low price, while the seller seeks a high price and is indifferent about the quality. If the buyer concedes regularly among the two issues he will offer to pay a low price for a high-quality TV, then an average price for a moderate-quality TV, and finally agree to a high price for a low-quality TV. The regular concession among issues, i.e. exploration of the outcome space with no diversity, made it impossible to agree on a high-quality TV for a high price, which is a better deal for the buyer and still acceptable for the seller.

Designing search algorithms (especially) with such specifications is challenging, mainly because of two reasons. First, the optimization problems associated with our three queries are hard to solve exactly and difficult to approximate rapidly. Second, guaranteeing all four specifications at once is difficult because often there are trade-offs between them. For instance, enumerating all possible outcomes of an outcome space, as in GeniusWEB [21], provides high accuracy and some diversity, but it is not scalable. Similarly, a search implemented through random sampling, as in NegMas [23], is scalable, diverse, and rapid, but it becomes really inaccurate as the number of negotiation issues increases.

3 Related Work

Most works in the fields of automated negotiations abstract away the outcome-space search method and assume there is an efficient way to implement it because the main topics of interest in field are the proposal of negotiation strategies and the design of negotiation protocols.

Jonker & Treur [18] propose the attribute-planning method, the earliest outcome-space search algorithm we are aware of. The authors in each round determine a utility target for the offer to be proposed and use it to define a utility target for each issue (additive utility functions are used). The method scales well and is rapid, however, it can have accuracy problem when applied to discrete issues and also the heuristic used to alter the target utility for each issue provides almost no diversity. The well known negotiation platform GENIUS [21] provides a default search method through which all possible outcomes are enumerated during a search process. The method provides high accuracy and some diversity, however, it does not scale well and it gets slow for moderately large outcome spaces. In NegMas [23] the proposed search method is based on random sampling and is therefore scalable and has high diversity, however because

Table 2. Comparison of search algorithms from literature with respect to the four design specifications, as well as their space & time complexity. In the table, $|\mathcal{I}|$ represents the number of negotiation issues, $|V|$ represents the highest number of possible values for one issue ($|V| = \max_i |V_i|$), and $|Im(d)|$ is the number of image points of the discretization operator d that BIDS uses (see Sect. 4.1) for more details.

	Scalable	Rapid	Accurate	Diverse	Space	Time
BIDS	✓	✓	✓	✓	$O(|\mathcal{I}| \cdot |Im(d)|)$	$O(1)$
Attribute-Planning [18]	✓	✓	?	×	$O(1)$	$O(|\mathcal{I}| \cdot |V|)$
Enumeration [21]	×	×	✓	✓	$O(|\mathcal{I}| \cdot |V|)$	$O(|V|^{|\mathcal{I}|})$
Random Sampling [23]	✓	✓	×	✓	$O(1)$	$O(1)$
MCTS [7]	✓	×	✓	✓	$O(|V|^{|\mathcal{I}|})$	$O(|V|^{|\mathcal{I}|})$
NB³ [7]	✓	×	✓	✓	$O(|V|^{|\mathcal{I}|})$	$O(|V|^{|\mathcal{I}|})$

it is tuned to be rapid it is not accurate on very large spaces. Participants of ANAC 2014 [3] were given the task to negotiate over large outcome spaces, under nonlinear preferences. Several meta-heuristics were proposed to implement the search, including simulated annealing [25] and genetic algorithm [11]. Similarly to random sampling, the proposed methods are scalable and provide high diversity, however, there is a trade-off in their tuning between being accurate and rapid. Buron et al. [7] propose a bidding strategy that uses Monte Carlo Tree Search [9] to explore the outcome space. Their method is heavily coupled with their negotiation strategy and while it can be scalable, accurate, and provide diversity, it is designed to operate under no time pressure.

In some other relevant works, Amini & Fathian [1] compare the performance of different stochastic search techniques in certain scenarios, with space sizes ranging from $59,049$ to $1,048,576$ possible outcomes, while de Jonge & Sierra [10] propose NB^3, a search algorithm in a setting where utility functions are publicly known, but computationally expensive (NP-hard) to calculate and there is no time pressure. Lastly, there is body of works which assumes dependencies between issues, represents them by graphs, and proposes negotiating strategies (and as a consequence search techniques) over them [16,17,22], however, the methods do not scale to the space sizes we are interested in.

4   Searching Through BIDS

We propose BIDS (**Bi**dding using **D**iversified **S**earch) — an algorithm that exploits the additive structure of a utility function to rapidly search very large outcome spaces providing accuracy & diversity. BIDS can answer the utility-lookup query, and it can also serve as a building block to tackle the utility-interval-sampling query and the trade-off query. To provide a tractable solution of the utility-lookup query, BIDS discretizes the codomain of the utility function and applies a dynamic-programming-based search to obtain an approximate solution to the associated optimization problem.

4.1   Looking for Bid(s) that Satisfy a Utility Target Through BIDS

A useful property of the utility-lookup query (see Eq. 1) is that its solution can be expressed through a recurrent relationship. Intuitively, if we suppose we know the solutions of Eq. 1 for $n-1$ issues and all possible utility thresholds no larger than u_t, then to solve the problem for n issues we have to simply pick the value of the nth issue that minimizes our objective function.

To formalize the idea we firstly need to consider partial outcomes: A *partial outcome* $\omega|_I$ is an outcome defined over only some issues $I \subset \mathcal{I}$, while Ω^P is the set of all partial outcomes over all possible subsets of issues. Furthermore, given a utility function $u : \Omega \to [0,1]$, we will denote by $u^P : \Omega^P \to [0,1]$ the extension of u over Ω^P:

$$u^P(\omega|_I) = \sum_{i \in I} \lambda_i \cdot u_i(\omega_i) \tag{4}$$

We define also the *concatenating operator* $+$ through which a value $v \in V_j$ for an issue $j \in \mathcal{I} \setminus I$ is attached to a partial outcome $\omega|_I = (\omega_1, \ldots, \omega_i)$:

$$\omega|_I + v_j = (\omega_1, \ldots, \omega_i, v_j). \tag{5}$$

Given this and denoting by $\sigma_k(u_t)$ the solution of Eq. 1 for a target utility u_t when the first k issues of Ω are used, the recurrent equation of utility-lookup query is:

$$\sigma_k(u_t) = \begin{cases} \operatorname{argmin}_{\omega_1 \in V_1} |u(\omega_1) - u_t|, & k = 1 \\ \operatorname{argmin}_{\omega_k \in V_k} u[\sigma_{k-1}(u_t - u^P(\omega_k)) + \omega_k], & \text{otherwise} \end{cases} \tag{6}$$

An algorithm that uses (6) to solve Eq. 1 will have exponential time complexity with respect to the number of negotiation issues. This is a result of the fact that the utility codomain is continuous and therefore the sub-problems created while solving the original problem are almost always non-overlapping. In other words, to provide a solution to recurrence (6), exponentially many sub-problems need to be solved.

Example 2 (Non-overlapping sub-problems). Suppose we want to find a bid close to utility target $u_t = 0.7$ in a negotiation over only three delivery dates of the example in Fig. 1. To calculate $\sigma_3(0.7)$, 365 sub-problems of calculating $\sigma_2(\cdot)$ need to be solved, each requiring yet another 365 partial solutions to $\sigma_1(\cdot)$. In general, since each issue-utility ranges over a continuous interval, there will be no overlap between the sub-problems that need to be solved, resulting in 365^3 calculations in the worst case.

The key to a tractable solution of Eq. 1 is to discretize the utility codomain and induce optimal sub-structure to the problem. BIDS does exactly this (see Algorithm 1) and as a consequence, can apply dynamic programming to calculate an approximate solution. To discretize the codomain while preventing negative utility thresholds from arising, BIDS uses the following discretization mapping:

$$d_p(u_t) = \begin{cases} \lfloor u_t \rceil_p, u \geq 0, \\ 0, otherwise \end{cases} \tag{7}$$

where $\lfloor \rceil_p : \mathbb{R} \to \mathbb{Q}$ rounds a real number at its pth decimal.

Algorithm 1. BIDS

Signature: BIDS$(u_t, I = \{1, \ldots, k\})$
1: **if** $k = 1$ **then**
2:     **return** $\mathrm{argmin}_{\omega_1 \in V_1} \quad |u^P(\omega_1) - u_t|$
3: **end if**
4: **return** $\mathrm{argmin}_{\omega_k \in V_k} u^P[\mathrm{BIDS}(d_p(u_t - u^P(\omega_k)), \{1, \ldots, k-1\}) + \omega_k]$

The table used by dynamic programming has $|\mathcal{I}| \cdot |Im(d)|$ entries, where $|\mathcal{I}|$ is the number of issues and $|Im(d)|$ is the number of image points of d (i.e. points in the grid defined over the utility codomain). As a consequence, the space computational complexity of BIDS is $O(|\mathcal{I}| \cdot |Im(d)|)$. Moreover, given that there are no more than $|V|$ possible values per issue, the time complexity of an implementation of BIDS that computes the dynamic programming table before the beginning of the negotiation and only searches the table in run-time is $O(|\mathcal{I}| \cdot |V| \cdot |Im(d)|)$ for the table-construction and $O(1)$ to search it during run-time. Lastly, there are trade-offs between the approximation accuracy of BIDS and its computational complexity.

Trading Computational Complexity for Approximation Accuracy. For simplicity, assume we use a regular grid over the utility codomain, with each point being 10^{-p} apart from its closest neighbors and where $p \in \mathbb{N}$ is the precision parameter which we can tune. Then the approximation error the method introduces in each iteration is at most 10^{-p}. Given that there are $|\mathcal{I}|$ issues, the algorithm runs $|\mathcal{I}|$ iterations for each solution. Therefore, the absolute error the method can introduce is $|\mathcal{I}| \cdot 10^{-p}$, which means the higher the precision, the smaller the introduced error. On the other hand, having grid points 10^{-p} apart implies that the grid is composed of 10^p points, which means the space complexity of the algorithm is $O(|\mathcal{I}| \cdot 10^p)$ and the time complexity of the table-construction is $O(|\mathcal{I}| \cdot |V| \cdot 10^p)$. Consequently, the more precise the algorithm is, the more space and construction time is going to require.

4.2 Using BIDS to Implement the Sampling-Utility Query and the Trade-off Query

BIDS can be used as a building block for algorithms that address the sampling-utility and the trade-off queries.

Algorithm 2 presents Sampling-BIDS, a method that provide n_s samples within a specified utility interval $\mathbb{I} = [u_{min}, u_{max}]$ in scalable, rapid, accurate, and diverse manner. The algorithm uses some arbitrary distribution — typically uniform — to sample n_s utility targets $U_t \subset \mathbb{I}$ (line 1 in the pseudo-code) and then uses BIDS to identify bids the utility of which is as close as possible to each of the targets. Its space complexity is the same as BIDS, i.e. $O(|\mathcal{I}| \cdot |Im(d)|)$, while the time complexity of an "offline" implementation is $O(n_s)$.

Algorithm 2. Sampling BIDS

Signature: Sampling-BIDS(n_s, $\mathcal{I}$, $[u_{min}, u_{max}]$)
1: $U_t := $ determineUtilSamples(n_s, u_{min}, u_{max})
2: $B := \emptyset$
3: **for** $u_t \in U_t$ **do**
4:     $B := B \cup \{\text{BIDS}(u_t, \mathcal{I})\}$
5: **end for**
6: **return** B

Similarly, Algorithm 3 presents Optimizing-BIDS, a method that builds upon Sampling-BIDS to approximately solve the trade-off query. Its space and time complexities are identical with Sampling-BIDS, i.e. $O(|\mathcal{I}| \cdot |Im(d)|)$ space complexity and $O(n_s)$ time complexity. Note that while Optimizing-BIDS improves the state-of-the-art by being scalable, rapid, and diverse, it does not provide accuracy guarantees.

Algorithm 3. Optimizing-BIDS

Signature: Optimizing-BIDS(n_s, $\mathcal{I}$, u_t)
1: $B := $ SamplingBIDS($n_s, \mathcal{I}, [u_t, 1.0]$)
2: **return** $\text{argmin}_{\omega \in B} f(\omega)$

---

5   Experiments

BIDS permits the implementation of the three most used search queries for outcome spaces. To validate the utility-lookup query, we have implemented it in GeniusWEB [21], so that state-of-art agents can use it. Furthermore, to verify that it complies to the four design specifications, we have designed two experiments: In Experiment 1 we investigate how scalable and rapid BIDS is by implementing a simple agent that uses the algorithm to explore the outcome space and compare it to various other agents. In Experiment 2 we isolate the search problem and compare BIDS with the scalable methods of the first experiment in terms of accuracy and diversity.

5.1 Setup

We run simulations for scenarios with arbitrary outcome spaces, containing 50 to 300 issues, with each issue having 10 possible values. We assign to each party an arbitrary utility profile over the generated spaces, i.e. an additive utility with random weights and random issue utilities.

First, we compare a simple agent that uses BIDS (BAgent) to all ANAC2021 participants. Since we are interested in the search process and not the negotiation strategy as a whole, in each round BAgent refuses the opponent's offer, sets an arbitrary utility target u_t, and uses BIDS with a precision $p = 5$ to identify the bid with utility as close as possible to u_t. In order to obtain an upper bound for scalability for a fixed negotiation, we allow each agent to use as much time as possible, by letting them negotiate against an opponent that uses minimal time to respond since it always rejects the opponent's offer and proposes a random bid as a counter-offer. Next, to obtain some accuracy and diversity results for very large outcome spaces, we compare BIDS to other scalable search algorithms. In particular, we compare it to attribute-planning [18], an adaptation of AgentM's [25] Simulated Annealing that answers the utility-lookup query, and GANGSTER's [11] Genetic Algorithm adapted to the utility-lookup query (for all three search methods we use the parameters proposed by their authors).

Lastly, note that we do not need any extensive computational resources. We run our simulations in a laptop with an i7 core and 16 GB of RAM.

5.2 Metrics to Quantify Scalability, Speed, Accuracy, and Diversity

Each search algorithm is scored on a number of metrics. To measure scalability, we count the highest number of issues for which an agent is able to exchange at least one offer. To also have a sense of how rapid each method is, we run negotiation sessions that last 2 min since ANAC2021 agents are designed to participate in negotiations of that length. Accuracy for the utility-lookup query is estimated by calculating the mean absolute error of the query's response from the defined target utility. More specifically, assume we pose queries for n different utility targets $U_t = \{u_{t_j}\}_{j=1}^n$ and get one response per each $\{\omega^1, \ldots, \omega^n\}$. Then we mean error, which we use to measure accuracy is:

$$e = \frac{1}{n} \sum_{j=1}^{n} |u_{t_j} - u(\omega^j)|$$

Lastly, to estimate diversity we quantify the change for two consecutive bids composition, i.e. we calculate the variability (through the standard error) of the concession rates among issues for two consecutive bids. More specifically, we define variability $v(\omega^j)$ of a bid ω^j with respect to its predecessor ω^{j-1} in the following way:

$$v(\omega^j) = \frac{1}{|\mathcal{I}|} \sum_{i=1}^{|\mathcal{I}|} [u_i(\omega_i^j) - u_i(\omega_i^{j-1})]$$

Then for a series bids $S = \{\omega^1, \ldots, \omega^n\}$ that answer queries we measure the series variability $v(S)$ as:

$$v(S) = SE(\{v(\omega^2), \ldots, v(\omega^n)\})$$

where SE stands for the standard error.

5.3   Experiment 1 - Scalability and Rapidness of BIDS

In the first experiment we investigate the scalability and rapidness of BIDS algorithm and compare it to ANAC2021 agents.

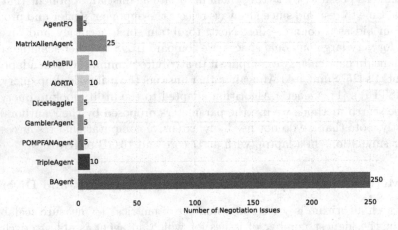

Fig. 3. Scalability results for BAgent and ANAC2021 participants.

Our results show that (see Fig. 3) BIDS can enable an agent to negotiate over 250 issues — over outcome spaces with 10^{250} possible outcomes — within a 2-minutes session, while the best performing ANAC2021 participant can negotiate upon a maximum of 25 issues — or over outcome spaces with 10^{25} possible outcomes. The poor performance of ANAC2021 comes as a result of their search method, with each agent using either: (a) exhaustive enumeration and cannot negotiate over more than 10 issues; or (b) some random sampling with no time constrains and are able to generate at least one offer for domains with up to 25 issues. Our BAgent cannot negotiate over more than 250 issues within 2 min since the initialization of the dynamic-programming table takes too long.

The results of Experiment 1 support our claim over the scalability of BIDS. However, if we focus solely on scalability, similar results can be achieved by letting BAgent use other search methods (see Fig. 4). Nonetheless, apart from scalability given some time restrictions, search accuracy and diversity play a crucial role in the quality of a search algorithm.

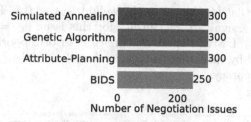

Fig. 4. Scalability results for BIDS, Attribute-Planning, Simulated Annealing, and Genetic Algorithm.

5.4 Experiment 2 - Accuracy and Diversity

In the second experiment we isolate the search problem to evaluate the search accuracy and diversity of BIDS algorithm and compare it with the other scalable methods (the attribute-planning, simulated annealing, and genetic algorithm) to get an insight of which algorithms can perform better in very large outcome spaces. To do so, we define a series of utility targets from 0 up to 1 with a regular step of 0.1 and use each search method to respond the utility-lookup query.

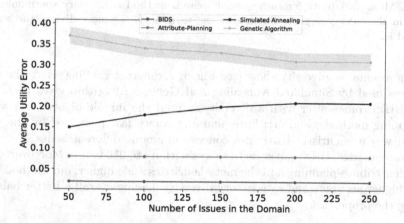

Fig. 5. Mean standard error for each scalable search method as we vary the number of issues in the outcome space. The smaller the error, the more accurate the search method is.

Our results on accuracy show (see Fig. 5) a clear ranking among the search methods. BIDS is more accurate since the way it explores the outcome space allows it to consider outcomes smartly and guarantee small error bounds (see Error Analysis in Sect. 4.1). Attribute-planning comes second penalized by the fact that it determines individual issue utility targets, which can lead to higher errors in discrete spaces. To illustrate this, suppose that in a given space for a particular issue ω_i there are only 2 possible values that can bring issue-utilities of 0.1 and 0.2 and that for that issue the weight is $\lambda_i = 0.5$. This means that if attribute-planning assigns a target utility for this issue $u_{t_i} = 0.8$, it will introduce an error of 0.3. Lastly, the meta-heuristics perform poorly with respect to accuracy, penalized by their randomness combined with their trade-off between search-time and accuracy.

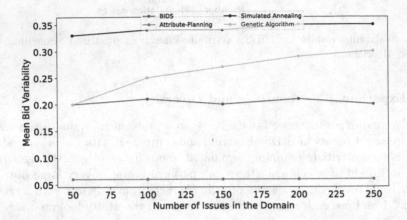

Fig. 6. Mean variability for each scalable search method as we vary the number of issues in the outcome space. The higher the variability, the more diverse the search method is.

The results on diversity show (see Fig. 6) a different ranking. Here the randomness used by Simulated Annealing and Genetic Algorithm gives them the lead; BIDS comes after with a diversity around the middle of best and worst performing methods; and attribute-planning comes last penalized by the very regular way its heuristic distributes concession among different issues.

To summarize, BIDS can scale to spaces with up to 250 issues. Moreover, even though attribute-planning and the meta-heuristics scale higher, our method provides higher accuracy and satisfactory diversity, having overall a better balance among the properties.

6   Conclusions and Future Work

This work presents BIDS — an algorithm that exploits the additive structure of the utility function to search very large outcome spaces while providing search accuracy and diversity. We find that BIDS can increase drastically the domain size in which negotiating agents can still function, while providing very high accuracy and significant outcome diversity. Therefore, BIDS algorithm can enable state-of-the-art (and future) agents to negotiate over very large realistic domains such as the ones found in procurement and supply-chain management.

Future work may build on this one to evaluate the robustness of specific negotiation strategies on the accuracy and diversity of the used search method, or investigate how strategies that do not consider the space size perform compared to strategies designed for large spaces. Moreover, the existing experimental setup can be extended to evaluate the performance of BIDS when used for the utility-sampling and the trade-off queries.

Acknowledgements. The research reported in this article is part of Vidi research project VI.Vidi.203.044, financed by the Dutch Research Council (NWO).

References

1. Amini, M., Fathian, M.: Optimizing bid search in large outcome spaces for automated multi-issue negotiations using meta-heuristic methods. Decis. Sci. Lett. **10**(1), 1–20 (2021)
2. An, B., Lesser, V.R., Irwin, D.E., Zink, M.: Automated negotiation with decommitment for dynamic resource allocation in cloud computing. In: AAMAS, vol. 10, pp. 981–988 (2010)
3. Aydogan, R., et al.: A baseline for non-linear bilateral negotiations: the full results of the agents competing in ANAC 2014 (2016)
4. Aydoğan, R., Fujita, K., Baarslag, T., Jonker, C.M., Ito, T.: ANAC 2017: repeated multilateral negotiation league. In: Ito, T., Zhang, M., Aydoğan, R. (eds.) ACAN 2018. SCI, vol. 905, pp. 101–115. Springer, Singapore (2021). https://doi.org/10.1007/978-981-15-5869-6_7
5. Baarslag, T., Hindriks, K., Jonker, C.: A tit for tat negotiation strategy for real-time bilateral negotiations. In: Complex Automated Negotiations: Theories, Models, and Software Competitions, pp. 229–233. Springer (2013). https://doi.org/10.1007/978-3-642-30737-9_18
6. Baarslag, T., Hindriks, K., Jonker, C., Kraus, S., Lin, R.: The first automated negotiating agents competition (ANAC 2010). In: New Trends in Agent-based Complex Automated Negotiations, pp. 113–135. Springer (2012). https://doi.org/10.1007/978-3-642-24696-8_7
7. Buron, C.L.R., Guessoum, Z., Ductor, S.: MCTS-based automated negotiation agent. In: Baldoni, M., Dastani, M., Liao, B., Sakurai, Y., Zalila Wenkstern, R. (eds.) PRIMA 2019. LNCS (LNAI), vol. 11873, pp. 186–201. Springer, Cham (2019). https://doi.org/10.1007/978-3-030-33792-6_12
8. Byde, A., Yearworth, M., Chen, K.Y., Bartolini, C.: Autona: a system for automated multiple 1–1 negotiation. In: EEE International Conference on E-Commerce 2003. CEC 2003, pp. 59–67. IEEE (2003)
9. Coulom, R.: Efficient selectivity and backup operators in monte-carlo tree search. In: van den Herik, H.J., Ciancarini, P., Donkers, H.H.L.M.J. (eds.) CG 2006. LNCS, vol. 4630, pp. 72–83. Springer, Heidelberg (2007). https://doi.org/10.1007/978-3-540-75538-8_7
10. De Jonge, D., Sierra, C.: NB^3: A multilateral negotiation algorithm for large, non-linear agreement spaces with limited time. Auton. Agent. Multi-agent Syst. **29**(5), 896–942 (2015). https://doi.org/10.1007/s10458-014-9271-3
11. de Jonge, D., Sierra, C.: GANGSTER: an automated negotiator applying genetic algorithms. In: Fukuta, N., Ito, T., Zhang, M., Fujita, K., Robu, V. (eds.) Recent Advances in Agent-based Complex Automated Negotiation. SCI, vol. 638, pp. 225–234. Springer, Cham (2016). https://doi.org/10.1007/978-3-319-30307-9_14

12. Faratin, P., Sierra, C., Jennings, N.R.: Using similarity criteria to make negotiation trade-offs. In: Proceedings Fourth International Conference on MultiAgent Systems, pp. 119–126. IEEE (2000)
13. Faratin, P., Sierra, C., Jennings, N.R.: Negotiation decision functions for autonomous agents. Robot. Auton. Syst. **24**(3–4), 159–182 (1998)
14. Fujita, K., Aydoğan, R., Baarslag, T., Hindriks, K., Ito, T., Jonker, C.: The sixth automated negotiating agents competition (ANAC 2015). In: Fujita, K., et al. (eds.) Modern Approaches to Agent-based Complex Automated Negotiation. SCI, vol. 674, pp. 139–151. Springer, Cham (2017). https://doi.org/10.1007/978-3-319-51563-2_9
15. Fujita, K., et al.: The second automated negotiating agents competition (ANAC 2011). In: Complex Automated Negotiations: Theories, Models, and Software Competitions, pp. 183–197. Springer (2013). https://doi.org/10.1007/978-3-642-30737-9_11
16. Hadfi, R., Ito, T.: Modeling complex nonlinear utility spaces using utility hypergraphs. In: Torra, V., Narukawa, Y., Endo, Y. (eds.) MDAI 2014. LNCS (LNAI), vol. 8825, pp. 14–25. Springer, Cham (2014). https://doi.org/10.1007/978-3-319-12054-6_2
17. Ito, T., Hattori, H., Klein, M.: Multi-issue negotiation protocol for agents: exploring nonlinear utility spaces. In: IJCAI, vol. 7, pp. 1347–1352 (2007)
18. Jonker, C.M., Treur, J.: An agent architecture for multi-attribute negotiation. In: International Joint Conference on Artificial Intelligence, vol. 17, pp. 1195–1201. LAWRENCE ERLBAUM ASSOCIATES LTD (2001)
19. Kawaguchi, S., Fujita, K., Ito, T.: Compromising strategy based on estimated maximum utility for automated negotiation agents competition (ANAC-10). In: Mehrotra, K.G., Mohan, C.K., Oh, J.C., Varshney, P.K., Ali, M. (eds.) IEA/AIE 2011. LNCS (LNAI), vol. 6704, pp. 501–510. Springer, Heidelberg (2011). https://doi.org/10.1007/978-3-642-21827-9_51
20. Koeman, V.J., Boon, K., van den Oever, J.Z., Dumitru-Guzu, M., Stanculescu, L.C.: The fawkes agent—the ANAC 2013 negotiation contest winner. In: Fujita, K., Ito, T., Zhang, M., Robu, V. (eds.) Next Frontier in Agent-based Complex Automated Negotiation. SCI, vol. 596, pp. 143–151. Springer, Tokyo (2015). https://doi.org/10.1007/978-4-431-55525-4_10
21. Lin, R., Kraus, S., Baarslag, T., Tykhonov, D., Hindriks, K., Jonker, C.M.: Genius: an integrated environment for supporting the design of generic automated negotiators. Comput. Intell. **30**(1), 48–70 (2014)
22. Marsa-Maestre, I., Klein, M., Jonker, C.M., Aydoğan, R.: From problems to protocols: towards a negotiation handbook. Decis. Support Syst. **60**, 39–54 (2014)
23. Mohammad, Y., Nakadai, S., Greenwald, A.: NegMAS: a platform for situated negotiations. In: Aydoğan, R., Ito, T., Moustafa, A., Otsuka, T., Zhang, M. (eds.) ACAN 2019. SCI, vol. 958, pp. 57–75. Springer, Singapore (2021). https://doi.org/10.1007/978-981-16-0471-3_4
24. Mohammad, Y., Viqueira, E.A., Ayerza, N.A., Greenwald, A., Nakadai, S., Morinaga, S.: Supply chain management world. In: Baldoni, M., Dastani, M., Liao, B., Sakurai, Y., Zalila Wenkstern, R. (eds.) PRIMA 2019. LNCS (LNAI), vol. 11873, pp. 153–169. Springer, Cham (2019). https://doi.org/10.1007/978-3-030-33792-6_10
25. Niimi, M., Ito, T.: AgentM. In: Fukuta, N., Ito, T., Zhang, M., Fujita, K., Robu, V. (eds.) Recent Advances in Agent-based Complex Automated Negotiation. SCI, vol. 638, pp. 235–240. Springer, Cham (2016). https://doi.org/10.1007/978-3-319-30307-9_15

26. Osborne, M.J., Rubinstein, A.: A Course in Game Theory. MIT press (1994)
27. TU Delft: GeniusWeb platform (2019). https://ii.tudelft.nl/GeniusWeb/technicians.html. Accessed 04 Jan 2022
28. Williams, C.R., Robu, V., Gerding, E.H., Jennings, N.R.: An overview of the results and insights from the third automated negotiating agents competition (ANAC2012). In: Marsa-Maestre, I., Lopez-Carmona, M.A., Ito, T., Zhang, M., Bai, Q., Fujita, K. (eds.) Novel Insights in Agent-based Complex Automated Negotiation. SCI, vol. 535, pp. 151–162. Springer, Tokyo (2014). https://doi.org/10.1007/978-4-431-54758-7_9

Purposeful Failures as a Form of Culturally-Appropriate Intelligent Disobedience During Human-Robot Social Interaction

Casey C. Bennett[1]([⊠])(iD) and Benjamin Weiss[2]

[1] Department of Intelligence Computing, Hanyang University, Seoul, Korea
cabennet@hanyang.ac.kr
[2] Quality and Usability Lab, Technische Universität, Berlin, Germany
benjamin.weiss@tu-berlin.de

Abstract. Human-robot interaction (HRI) can suffer from *breakdowns* that are often regarded as "failures" by roboticists. Here, however, we argue that such breakdowns can be sometimes perceived as a type of *defiance* that signals more socially intelligent behavior rather than less, depending on the culture and linguistic environment within which they occur. We present recent research evidence supporting this viewpoint, based on HRI experiments comparing English speakers and Korean speakers. Counterintuitively, occasional culturally-appropriate forms of disobedience may in fact be a desirable design feature for social robots in the future.

Keywords: Human robot interaction · Failures · Culture · Social interaction · Inhibition of return · Autonomous agents

1 Introduction

An idea gaining broader attention recently is the concept of social interaction breakdown during human-robot interaction (HRI) studies. Within the HRI field, these breakdowns are often regarded as "failures" [2,18]. While on the surface that may be true in principle, there is a growing body of HRI research that focuses on cross-cultural robotics, with the aim of conducting the exact same experiment in multiple geographic locations as well as multiple languages/cultures [3,30,31]. This involves both physically embodied robots as well as virtual avatars. Results from that research have led us to an interesting notion, that **some breakdowns may not be regarded as failures at all in certain cultural environments, but rather may be perceived as more intelligent behavior in the form of occasional defiance of existing social norms**. Moreover, what may be viewed as "normal" behavior in one

This work was supported through funding by a grant from the National Research Foundation of Korea (NRF grant# 2021R1G1A1003801).

© Springer Nature Switzerland AG 2022
F. S. Melo and F. Fang (Eds.): AAMAS 2022 Workshops, LNAI 13441, pp. 84–90, 2022.
https://doi.org/10.1007/978-3-031-20179-0_5

cultural environment may be seen as "defiant" in another [9]. In that sense, such *perceived disobedience* may in fact create an opportunity to design more effective social robots.

A good example of this is a comparison between East Asian and Western cultures, for which there is significant existing research [21]. In particular, what constitutes appropriate robot/agent behavior in those environments is defined by several dimensions: high-context vs. low-context cultures, communal vs. individualistic cultures, Confucian power hierarchies vs. Western power hierarchies, etc. [5,27,30]. A broad review of such existing research can be found in Lim et al. 2021 [24].

Such cultural differences are also embedded into our spoken languages, otherwise known as *linguistic relativity* [13,15]. The basic argument is that the language we speak affects our habitual behaviors and worldview [8], much like a lens warps visual perception. In bilingual speakers, this can constitute cognitive differences depending on the language they are actively speaking in [22]. This lens metaphor extends to expectations, e.g. the right amount of gaze or verbal backchannel depends on (culturally different) roles and hierarchies [20]. If associated behavior does not meet the expectation, it can irritate a dialogue partner so much that it requires explicit clarification, disrupting the flow of communication [14]. This, however, is culturally dependent, e.g. a strong indicator of a breakdown in some Western cultures, silence, can represent a very appropriate contribution to dialogue in high-context cultures [10].

Below we present a brief summary of some recently completed research related to these topics and this workshop on rebellion and disobedience in AI [12,25]. We then discuss our perspectives on how research around cross-cultural robotics can contribute to the broader discussion of intelligent disobedience and its applications within the HRI field.

2 Methods

Our own research lends further evidence to this position. In particular, one of our recently concluded studies focuses on the concept of *Social Inhibition of Return* (social IOR), which is based on IOR models from various human sensory functions such as vision [26]. The basic idea here is that there are mechanisms in the brains of naturally intelligent organisms (including humans) that inhibit us from repeating the same behavior in a short period of time (e.g. 2–3 s) in order to maximize task efficiency (e.g. during visual "information foraging") [19]. A failure in these mechanisms is thought to play a role in human mental illness, such as obsessive-compulsive disorder. These mechanisms are also important to produce fluid natural behavior, rather than repetitive "robot-like behavior" in humans [26]. In an HRI context, the removal of social IOR causes the agent to engage in repetitive speech behaviors as well as increase the chance of "talking over" the human interactor. **Such behavior theoretically could be seen as *defying* typical social norms, or perhaps even "aggressive" social behavior by the agent**, rather than trying to engage the human on their terms (as is the typical approach in HRI). We posit that such defiance may affect perceived disobedience of the robot by the human, as defined in Sect. 1.

In the recent study, we experimented with social IOR, by having a virtual avatar play a social survival video game called "Don't Starve Together" with a human player. The agent was capable of autonomous speech interactions during gameplay (i.e. *Social AI*), which was developed specifically for cooperative game paradigms during previous studies [6]. This specific Social AI was capable of hundreds of different speech utterances covering 46 different utterance categories, each related to a particular game situation (e.g. collecting resources, fighting monsters, deciding where to go next) organized as a hierarchy with several levels. Those speech utterances were both self-generated based on internal logic of the Social AI, as well as responses to human player speech via automatic speech recognition (ASR). The speech responses were similar in both English and Korean (i.e. the robot was bilingual, in essence).

Fig. 1. Gameplay example during the experiment (human vs avatar)

In order to implement the social IOR in our case, we utilized the top-level utterance categories from the speech hierarchy (6 total) so that the Social AI maintained an internal array to keep track of recently spoken categories, with a "counter" that counted down a certain number of seconds during which any further utterances within that same category were suppressed (though the AI could still make utterances from other categories). This counter was set to 3 s, based on prior research on social IOR in humans [26]. The study described here involved a control condition, including the social·IOR, and an experimental condition where the social IOR was turned off. The total sample size was 32 participants (16 in each condition), with 16 Korean speakers and 16 English speakers split into each condition. Participants played the game with the virtual avatar for 30 min total (see Fig. 1). To evaluate the effects, we utilized two common standardized scales: Godspeed scale [1] for measuring perceptions of a robot/agent and the Networked Minds instrument [7] for measuring *social presence* [28].

3   Results

Our initial hypothesis was that social IOR would enhance human perception of the interaction with the virtual avatar agent. However, results suggest the effect may be dependent on the language of the speaker. While there was minimal change in the total *social presence* values as per the Networked Minds instrument in Korean speakers without social IOR vs with (0.80 vs 0.85), there was a significant effect in English speakers (0.91 vs. 0.37), as seen in Table 1. This was largely due to increases in both *attentional engagement* (the feeling that when I pay attention to something, the agent does too) and *emotional contagion* (the feeling that when I feel something, the agent feels that way too) in English speakers. Conversely, Korean speakers appeared to have either no change or a slight reduction in those dimensions when social IOR was removed.

Table 1. Networked minds - social presence

Language	With sIOR	No sIOR
Korean	0.85	0.80
English	0.37	0.91

Table 2. Godspeed likeability

Language	With sIOR	No sIOR
Korean	3.24	3.29
English	3.51	3.06

Likewise, the Godspeed scores were notably increased on average for English speakers (3.51 vs. 3.06) but not the Korean speakers (3.24 vs. 3.29), as seen in Table 2. In short, whether removing social IOR seems to make the agent more engaging, but less likeable, appears to be dependent on the language of the speaker. We discuss the design implications of this for socially interactive agents further in the next section.

4   Discussion

Social interaction is a notoriously amorphous domain, where it is not always clear how to define the "goals" an agent should pursue [29] and determining the outcome of such goals may be subjective in nature [16]. That leads to challenges with planning for agent behavior, as well as rational decision-making for AI systems in social situations. We contend here that *culturally-appropriate* defiance of social norms can counterintuitively help create autonomous agents and robots that are perceived as more socially intelligent.

This is in line with suggestions from previous research [12,25], though perhaps more similar to the latter's "intelligent disobedience" approach than the former. In our case, disobedience is not triggered by any specific factors but rather is a design consideration for creating more seemingly intelligent behavior (similar to [17]). **Indeed, a lack of such occasional disobedience may also explain past results in cross-cultural robotics**, which indicate that cultural homophily (e.g. agents adapted to a specific set of cultural attributes) alone does not necessarily correspond to higher ratings of a robot by humans [24]. In certain cultural settings, sporadic purposeful "failures" may actually be a desirable design feature.

Much previous research in HRI has argued for "culturally-robust" or "culturally-aware" systems (including our own), where robots are designed specifically to create adaptable behaviors that match the value system of the local human culture [11,23,24,30]. While that is certainly one approach for *value alignment* in social robots, here we take the position that it may be necessary to create different models (machine learning or otherwise) of robot behavior specifically for different cultures, taking into account differential responses to perceived disobedience in the robot by human interactors. Similar to the current research evidence provided in this paper, we are seeing the same phenomenon in another study involving robotic pets placed into user homes in Asia and the United States [4]. This topic of perceived disobedience towards better social intelligence is an area of rich potential, as there are many ways to violate social norms, which can be dependent on both behavior (speech, facial expressions, gesture) as well as robotic form factor.

References

1. Bartneck, C., Kulić, D., Croft, E., Zoghbi, S.: Measurement instruments for the anthropomorphism, animacy, likeability, perceived intelligence, and perceived safety of robots. Int. J. Soc. Robot. 1(1), 71–81 (2009). https://doi.org/10.1007/s12369-008-0001-3
2. Bennett, C.C.: Evoking an intentional stance during human-agent social interaction: appearances can be deceiving. In: IEEE International Symposium on Robot and Human interactive Communication (RO-MAN), pp. 362–368 (2021). https://doi.org/10.1109/RO-MAN50785.2021.9515420
3. Bennett, C.C., Sabanovic, S., Fraune, M.R., Shaw, K.: Context congruency and robotic facial expressions: do effects on human perceptions vary across culture? In:

IEEE International Symposium on Robot and Human interactive Communication (RO-MAN), pp. 465–470 (2014)

4. Bennett, C.C., et al.: Comparison of in-home robotic companion pet use in south korea and the united states: a case study. In: 9th IEEE International Conference on Biomedical Robotics & Biomechatronics (BIOROB), In Press (2022)

5. Bennett, C.C., Stanojevic, C., Sabanovic, S., Piatt, J.A., Kim, S.: When no one is watching: ecological momentary assessment to understand situated social robot use in healthcare. In: 9th International Conference on Human-Agent Interaction (HAI), pp. 245–251 (2021). https://doi.org/10.1145/3472307.3484670

6. Bennett, C.C., Weiss, B., Suh, J., Yoon, E., Jeong, J., Chae, Y.: Exploring data-driven components of socially intelligent ai through cooperative game paradigms. Multimodal Technol. Interact. **6**(2), 16 (2022). https://doi.org/10.3390/mti6020016

7. Biocca, F., Harms, C., Gregg, J.: Does language shape thought?: Mandarin and english speakers' conceptions of time. In: 4th Annual International Workshop on Presence, pp. 1–9 (2001). https://doi.org/10.1006/cogp.2001.0748

8. Boroditsky, L.: Does language shape thought?: mandarin and english speakers' conceptions of time. Cogn. Psychol. **43**(1), 1–22 (2001). https://doi.org/10.1006/cogp.2001.0748

9. Briggs, G., Williams, T., Jackson, R.B., Scheutz, M.: Why and how robots should say 'No'. Int. J. Soc. Robot. **14**(2), 323–339 (2021). https://doi.org/10.1007/s12369-021-00780-y

10. Bruneau, T.J.: Communicative silences: forms and functions. J. Commun. **23**(1), 17–46 (1973). https://doi.org/10.1111/j.1460-2466.1973.tb00929.x

11. Bruno, B., Menicatti, R., Recchiuto, C.T., Lagrue, E., Pandey, A.K., Sgorbissa, A.: Culturally-competent human-robot verbal interaction. In: 5th International Conference on Ubiquitous Robots (UR), pp. 388–395 (2018). https://doi.org/10.1109/URAI.2018.8442208

12. Coman, A., Aha, D.W.: Ai rebel agents. AI Mag. **39**(3), 16–26 (2018). https://doi.org/10.1609/aimag.v39i3.2762

13. Deutscher, G.: Through the language glass: why the world looks different in other languages. Metropolitan Books (2010)

14. Enfield, N.J.: How we talk: the inner workings of conversation. In: Basic Books (2017)

15. Fuhrman, O., Boroditsky, L.: Cross-cultural differences in mental representations of time: evidence from an implicit nonlinguistic task. Cogn. Sci. **34**(8), 1430–1451 (2010). https://doi.org/10.1111/j.1551-6709.2010.01105.x

16. Gordon, G.: Infant-inspired intrinsically motivated curious robots. Curr. Opin. Behav. Sci. **35**, 28–34 (2020). https://doi.org/10.1016/j.cobeha.2020.05.010

17. Hiatt, L.M., Harrison, A.M., Trafton, J.G.: Accommodating human variability in human-robot teams through theory of mind. In: Twenty-Second International Joint Conference on Artificial Intelligence (IJCAI) (2011). https://doi.org/10.5555/2283696.2283745

18. Honig, S., Oron-Gilad, T.: Understanding and resolving failures in human-robot interaction: literature review and model development. Front. Psychol. **9**, 861 (2018). https://doi.org/10.3389/fpsyg.2018.00861

19. Klein, R.M., MacInnes, W.J.: Inhibition of return is a foraging facilitator in visual search. Psychol. Sci. **10**(4), 346–352 (1999). https://doi.org/10.1111/1467-9280.00166

20. Knapp, M.L., Hall, J.A., Horgan, T.G.: Nonverbal communication in human interaction. In: Cengage Learning (2013)

21. Komatsu, T., Malle, B.F., Scheutz, M.: Blaming the reluctant robot: parallel blame judgments for robots in moral dilemmas across us and Japan. In: ACM/IEEE International Conference on Human-Robot Interaction (HRI), vol. 9, pp. 63–72 (2021). https://doi.org/10.1145/3434073.3444672
22. Kousta, S.T., Vinson, D.P., Vigliocco, G.: Investigating linguistic relativity through bilingualism: the case of grammatical gender. J. Exp. Psychol. Learn. Mem. Cogn. **34**(4), 843 (2008). https://doi.org/10.1037/0278-7393.34.4.843
23. Lee, H.R., Sabanovic, S.: Culturally variable preferences for robot design and use in South Korea, Turkey, and the United States. In: ACM/IEEE International Conference on Human-Robot Interaction (HRI), pp. 17–24 (2014). https://doi.org/10.1145/2559636.2559676
24. Lim, V., Rooksby, M., Cross, E.S.: Social robots on a global stage: establishing a role for culture during human–robot interaction. Int. J. Soc. Robot. **13**(6), 1307–1333 (2020). https://doi.org/10.1007/s12369-020-00710-4
25. Mirsky, R., Stone, P.: The seeing-eye robot grand challenge: rethinking automated care. In: 20th International Conference on Autonomous Agents and MultiAgent Systems, pp. 28–33 (2021). https://doi.org/10.5555/3463952.3463959
26. Nafcha, O., Shamay-Tsoory, S., Gabay, S.: The sociality of social inhibition of return. Cognition **195**, 104108 (2020). https://doi.org/10.1016/j.cognition.2019.104108
27. Nomura, T.T., Syrdal, D.S., Dautenhahn, K.: Differences on social acceptance of humanoid robots between japan and the Uk. In: Proceedings of the 4th International Symposium on New Frontiers in Human-Robot Interaction, The Society for the Study of Artificial Intelligence and the Simulation of Behaviour (AISB) (2015)
28. Oh, C.S., Bailenson, J.N., Welch, G.F.: A systematic review of social presence: definition, antecedents, and implications. Front. Robot. AI **5**, 114 (2018). https://doi.org/10.3389/frobt.2018.00114
29. Rolf, M., Crook, N.T.: What if: Robots create novel goals? ethics based on social value systems. In: EDIA Workshop at the European Conference on Artificial Intelligence (ECAI), pp. 20–25 (2016)
30. Sabanovic, S., Bennett, C.C., Lee, H.R.: Towards culturally robust robots: a critical social perspective on robotics and culture. In: Proceedings of the HRI Workshop on Culture-Aware Robotics (2014)
31. Weiss, B., Wechsung, I., Kühnel, C., Möller, S.: Evaluating embodied conversational agents in multimodal interfaces. Comput. Cogn. Sci. **1**(1), 1–21 (2015). https://doi.org/10.1186/s40469-015-0006-9

Visionary Papers

An Agent-Based Model of Horizontal Mergers

Martin Harry Vargas Barrenechea🆔 and José Bruno do Nascimento Clementino(✉)🆔

Federal University of Ouro Preto, Mariana, MG, Brazil
mbarrenecha@ufop.edu.br, jose.clementino@aluno.ufop.edu.br

Abstract. This paper is about the agentization of a horizontal mergers model. In this model, firms are either in a differentiated products Bertrand competition, in which they choose prices in order to maximize their profits, or in a Cournot competition, in which quantities are chosen by firms. The analytical game theoretical model predicts that once a firm merges to another, prices of the merging party rise, which leads to a decrease in consumer surplus and an increase in producer surplus. Developed along this draft is an agent-based version of this model in which firms do not know the demand they are facing. We find convergence of our agent-based model to the game theoretical results before and after firms merge. Alternative learning methods will be implemented as a further extension to this agent-based model.

Keywords: Horizontal Mergers · Agent-based models · Game Theory

1 Introduction

A horizontal merger occurs when competitors in an industry merge. Since this event leads to the concentration of an industry, one of its possible consequences is the exercise of unilateral market power – which could translate into a raise in prices not only of the goods involved in merging parties, but of the whole market. According to the Federal Trade Commission, the regulation agency responsible for evaluating mergers in the United States, over a thousand merger cases are reviewed every year[1]. In Europe, the European Commission (EC) received around 230 notifications until july of this year[2]. As stated by the Federal Trade Commission, around 5% of the cases reviewed by the agency present competitive issues. In that case, to prevent mergers that are damaging to the consumer's welfare and competition, it is necessary to study and develop methods that help verifying if mergers will affect competition negatively.

A significant amount of methodologies that measure the effects of mergers have been developed such as [1–3,5]. All of these methods use analytical and statistical tools for the estimation of prices before and after mergers. However, there's an absence of models that describe and reproduce such effects of unilateral market power in the agent-based computational economics literature. Although industrial organization studies the strategic interaction between firms, which could be represented through agent-based

---

[1] See: https://www.ftc.gov/tips-advice/competition-guidance/guide-antitrust-laws/mergers.
[2] See:https://ec.europa.eu/competition-policy/mergers/statistics_en.

© Springer Nature Switzerland AG 2022
F. S. Melo and F. Fang (Eds.): AAMAS 2022 Workshops, LNAI 13441, pp. 93–105, 2022.
https://doi.org/10.1007/978-3-031-20179-0_6

modeling tools, [11] states a lack of integration between the industrial organization theory and agent-based methodologies. This paper is a small contribution to narrow the gap between these two research areas.

Using a constructive approach, as proposed by [14], an agent-based model of horizontal mergers is developed. A model presented in [10], where firms are in a differentiated products Bertrand game, is used as a benchmark case and goes through the process of *agentization*, which is also described in [6]. The model is also extended to the Cournot game. Assumptions of perfect rationality and information are relaxed from the model, and the emerging patterns before and after mergers are studied. Under our chosen assumptions of agent learning, qualitative and quantitative results are identical to the analytical model.

The work is divided into four sections. In the first section, the exposition and assumptions of the analytical model are presented. The second section presents the agent-based version of this model. Presented in the third section are the results obtained from simulations. Finally, the fourth has concluding remarks about this work and extensions that we plan to include on a future version.

2   The Analytical Model

As mentioned earlier, this model is largely based on the model of horizontal mergers featured in [10, p.244–265]. To the reader that might not be familiar with the concept of a horizontal merger, these mergers happen whenever firms of the same industry merge[3]. A quick exposition of the model is going to be done in this section, showing the utility of a representative consumer and how it determines quantities given the prices of the goods in the market and how firms, considering this demand, determine optimal prices.

2.1   Bertrand Competition with Differentiated Products

The utility of the representative consumer depends on multiple differentiated products in the market. It is represented as:

$$U = v \sum_{i=1}^{n} q_i - \frac{n}{2+\gamma} \left[\sum_{i=1}^{n} q_i^2 + \frac{\gamma}{n} \left(\sum_{i=1}^{n} q_i \right)^2 \right] + y \tag{1}$$

where y is an outside good, and since this demand is quasi-linear, it does not affect the decisions taken by the consumer with respect to the differentiated products; q_i is the quantity of the i-th product; v is a positive parameter; n is the number of products in the industry; and γ represents the degree of substitutability between the n products. From utility maximization, prices and quantities can be determined. The direct demand function is defined as:

$$q_i = \frac{1}{n} \left[v - p_i(1 + \gamma) + \frac{\gamma}{n} \sum_{j=1}^{n} p_j \right] \tag{2}$$

---

[3] One alternative would be the occasion in which a firm sells input to another firm that produces a final good. In the case that one firm merges with the other, this event would be called a vertical merger.

Quick inspection shows that the demand for the i-th good depends on its own price and the price of other goods, that is, if a good has a price that is too high, it will be substituted by other products, depending on the degree of substitutability, which will lead to a diminished demand for this product. However, if other goods are too expensive, then the good in question will be purchased abundantly. Another property of this demand function is the fact that the aggregate demand does not depend on the degree of substitution among the products. Finally, if prices are identical among all firms, the aggregate quantity does not change with the number of products that exist in an industry.

This model presumes that each firm in an industry sells a single good. The profit attained from selling each good is defined as:

$$\pi_i = (p_i - c)(q_i) \tag{3}$$

where c is the marginal cost for producing a unit of the i-th good. For simplicity, this model presumes that there are no fixed costs in the cost function and costs are homogeneous among firms. The firm's expected behavior is to choose a price that maximizes its profit. By substituting Eq. 2 on Eq. 3, taking the derivative with respect to price and setting it to zero, the game theoretical price for the i-th good will be:

$$p_i = \frac{nv + \gamma \sum_{j=1, j\neq i}^{n} p_j + c(n + n\gamma - \gamma)}{2(n + n\gamma - \gamma)} \tag{4}$$

When a firm merges to other firms, the merged party turns into a multi-product firm. This new firm sells m products while the remaining firms sell $m - n$ products. The profit of the multi-product firm will be the sum of the attained profits when considering all goods involved in the merge. Game theoretical prices are obtained by taking the derivative of both products and setting it to zero, the results will be:

$$p_I(m) = \frac{c(n\gamma(4n - 2m - 1) + 2n^2 + \gamma^2(2n^2 - nm - 2n - m^2 + 2m)) + nv(2n + \gamma(2n - 1))}{\gamma^2(2n^2 - nm - 2n - m^2 + 2m) + 2\gamma n(3n - m - 1) + 4n^2} \tag{5}$$

$$p_o(m) = \frac{c(n\gamma(4n - m - 2) + 2n^2 + \gamma^2(2n^2 - nm - 2n - m^2 + 2m)) + nv(2n + \gamma(2n - m))}{\gamma^2(2n^2 - nm - 2n - m^2 + 2m) + 2\gamma n(3n - m - 1) + 4n^2} \tag{6}$$

where Eq. 5 is the price of the multi-product firm and Eq. 6 is the price of the outside firm (that is, the firm that is not in the merging party). Notice how both profits are in terms of the products sold by the merging party. The higher the number of products sold by the multi-product firm, the higher are profits for both types of firms. This suggests that the consumer surplus decreases in the presence of a merger[4].

2.2 Cournot Competition with Differentiated Products

In this paper, we also consider the agentization of the Cournot competition with differentiated products model. The utility of the representative agent is identical to the

---

[4] To the reader that is unfamiliar with the consumer surplus, it is described as the amount of utility obtained by a consumer after a transaction. A simple formula for it would be $CS = \sum_{i=1}^{n} U(q_i) - q \cdot p$, where q and p are vectors for quantities and prices respectively. With quantities held constant, an increase in prices leads to a decrease in CS.

Bertrand case with the exception that instead of inserting Eq. 2 into the profit function, the indirect demand is inserted. This indirect demand is given by the following equation:

$$p_i = v - \frac{1}{1+\gamma}(nq_i + \gamma \sum_{j=1}^{n} q_j) \tag{7}$$

In this model, the firm's expected behavior is to choose a quantity that maximizes its profit. The game theoretical quantity for the i-th good will be given by the following equation:

$$q_i = \frac{(v-c)(1+\gamma) - \gamma \sum_{j=1, j \neq i}^{n} q_j}{2(n+\gamma)} \tag{8}$$

Like the Bertrand case, we consider that a firm merges to other firms, which leads to the creation of a multi-product firm that sells m products. The remaining firms sell $(n-m)$ products. The game theoretical quantities that are obtained from this new market arrangement are given by the following equations:

$$q_I(m) = \frac{(v-c)(1+\gamma)(2n+\gamma)}{(2n+2m\gamma)(2n+\gamma(n-m+1)) - mn\gamma^2 + \gamma^2 m^2} \tag{9}$$

$$q_o(m) = \frac{(v-c)(1+\gamma)[(2n+2m\gamma)(2n+\gamma(n-m+1)) - mn\gamma^2 + \gamma^2 m^2 - 2\gamma(2n+\gamma)]}{(2n+\gamma(n-m+1))((2n+2m\gamma)(2n+\gamma(n-m+1)) - mn\gamma^2 + \gamma^2 m^2)} \tag{10}$$

where Eq. 9 is the quantity sold of each product in the multi-product firm portfolio and Eq. 10 is the quantity sold by firms that are outside of the merging party. These quantities are conceptually similar to the Bertrand case, because it assumes that the firms are in an symmetric equilibrium. However, mergers under a Cournot competition are not necessarily profitable. This happens because merged firms produce a smaller quantity than non-merged firms, and unless products are extremely differentiated, which is associated to a small γ, the firms that are outside the merger will compensate the smaller production from the multi-product firm in order to increase their own profits.

3   The Agent-Based Model

The last section gave a quick overview regarding the analytical model of mergers. Describing the agentization of the model is the purpose of this section. As is going to be shown through the Overview, Design Concepts, and Details (ODD) protocol based on [13], the agentization of this model is going to occur with respect to firm behavior. Instead of firms that have access to perfect information and rationality, firms are rationally bounded. With this we are effectively relaxing some of the hypotheses of the original model.

Purpose and Patterns

The purpose of the model is to reproduce the game theoretical results of a differentiated products Bertrand competition. Mergers lead to market concentration, which leads to a general increase in prices.

Entities, State Variables and Scales

The entities of the model are firms that are engaged in the competition. In this version of the model, geographic space is not relevant. Simulations are run from 1500 to 4000 periods. Every firm has the following state variables:

– *current price*: from which it draws price bids every period in the Bertrand case;
– *current quantity*: from which it draws quantity bids every period in the Cournot case;
– *cost*: the cost of producing a single good.

Process Overview and Scheduling

In the Bertrand case, at every period, firms draw price bids which determine quantities. After these quantities are determined, profits are calculated and saved on their memories. When prices are stable, firms engage in the merging process and a new price adjustment phase begins. The Cournot case is quite similar to the Bertrand one, the difference is regarding to the bids that are drawn by firms: instead of drawing price bids, firms draw quantity bids. For the sake of simplicity, we focus on the explanation of the Bertrand case.

Design Concepts

Firms try to learn prices that maximize their profits adaptively. Since firms are not able to see information from their competitors, the interaction is only through indirect means because their prices affect demand. From the learning process and adaptation, game theoretical prices emerge.

Initialization

In this version, firms are identical when considering current prices, costs and learning parameters. The only form of heterogeneity in this model is driven by stochasticity, because bids are chosen randomly, and it's not unlikely that bids are distinct from each other. Chosen values for parameters are presented in the following section.

Submodels

Besides the equation that is responsible for determining the demand, another submodel is defined for firm behavior. The learning method is based on [8]. It is an adaptation to a line search method which is useful for finding the optima of functions that have a single variable[5]. Firms draw bids from a uniform distribution:

$$bid \sim U(\text{current price} - \delta, \text{current price} + \delta) \tag{11}$$

after their bids are drawn, quantities are determined by Eq. 2. The results from their profits are saved into one of two lists: the first one is for when a firm bids higher than

---

[5] See [12] for an overview of the method.

its current price; the other list is for when a firm bids lower than its current price. After an "epoch", the name given for a learning phase consisting of 30 periods, the firm compares the mean from both lists. If the list related to high prices has a mean profit higher than the list of low prices, then:

$$new\ current\ price = current\ price + \epsilon \qquad (12)$$

else:

$$new\ current\ price = current\ price - \epsilon \qquad (13)$$

after the firm is done, the lists are emptied and a new learning phase begins. A simple pseudocode, adapted from [8, p.189] is presented for the sake of clarity on algorithm 1.

Another simple submodel for price stability (or equilibrium) is necessary for the initialization of a merger. Stability is defined in terms of moving averages. A moving average considers the current price of three epochs. Let μ_1 be the average of three epochs; after that, μ_2 is defined as the average of the next three epochs. If the absolute value of the difference between these two averages is lower than a threshold, that is $|\mu_2 - \mu_1| \le \theta$, then prices are stable, which means firms have found an equilibrium. If that is not the case, $\mu_1 = \mu_2$ and a new value of μ_2 is calculated considering the next three epochs.

Because we are interested in the means of current prices before and after mergers, mergers occur only after prices are stable. When two firms merge, the profit of each product is calculated individually, but the firms that are part of the merging party see their profit as the sum of their individual profits.

3.1 Model Implementation and Parameters

The model was implemented using Netlogo v6.2 ([15]). Its parameters used in simulations were the ones given in Table 1.

Under such parameters, the game theoretical (optimal price) is: 11.54, and the optimal quantity associated with that price is: 29.49. After a merger happens, the optimal price for the merging party is: 16.87; for the non-merging firms, the price will be 13.86. For the Cournot case, the Nash equilibrium quantity is 21.43, while price will be 35.71. After a merger happens, the quantity of the merging party will be 17.67, while the non-merged party will produce 28.71. Prices of the merged and non-merged parties will be very close, being 3.69 and 3.39 respectively.

In the Netlogo model, there are sliders that determine the value of learning parameters, such as ϵ and δ, demand parameters, such as γ and v, initial prices, firm costs, the number of firms (Fig. 1).

Algorithm 1. Probe and Adjust

1: Set learning parameters: $\delta, \epsilon, epoch\_length$
2: $counter \leftarrow 0$
3: $returns\_up \leftarrow []$ (An empty list associated with current price raises)
4: $returns\_down \leftarrow []$ (An empty list associated with current price decreases)
5: Do forever:
6:      $counter \leftarrow counter + 1$
7:      $price\_bid \sim U(current\_price - \delta, current\_price + \delta)$
8:      $profit \leftarrow$ Return of price_bid
9: **if** $bid\_price \geq current\_price$ **then**
10:         Append profit to returns_up
11: **else**
12:         Append profit to returns_down
13: **end if**
14: **if** ($counter$ mod $epoch\_length = 0$) **then** (This means the learning period is over.)
15:     **if** $mean\ returns\_up \geq mean\ returns\_down$ **then**
16:             $current\_price \leftarrow current\_price + \epsilon$
17:             $returns\_up \leftarrow []$
18:             $returns\_down \leftarrow []$
19:     **else**
20:             $current\_price \leftarrow current\_price - \epsilon$
21:     **end if**
22: **end if**
23: Back to step 5

Table 1. Parameters used in simulation

	Value
γ	10
v	100
ϵ	0.7
δ	3
θ	3
Epoch length	30
Initial price	20
Firm costs	0
Number of firms	3

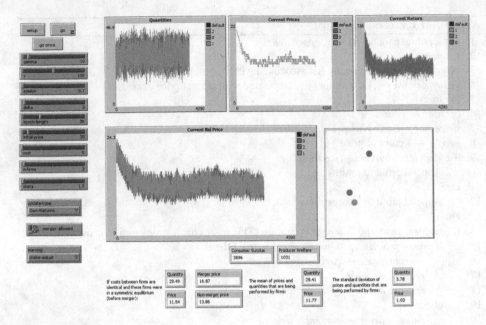

Fig. 1. This figure shows a typical run of the Netlogo based model.

4   Results

In this section we will discuss the results of the model considering the parameters chosen for our simulations. First we will analyze results obtained before mergers and then after mergers. Finally, statistical tests will be run to compare means of current prices between each of the circumstances.

Before Mergers

As a first experiment, 10 runs with the given parameters were considered to understand the model's behavior. Every period (step), the mean and standard deviation of prices, quantities and welfare values were taken. The data was grouped by step, so notice that these are the average of means and standard deviations. Figure 2 shows the progression of prices and quantities as time evolves. In our simulations, the stability of prices occur around the 600*th* period.

At the optimum, considering symmetric prices, quantities should be around 29.49. The mean quantity produced by firms follows that very closely, as seen on the right panel of the Fig. 2. The mean current quantity is around 29.4 with a standard deviation of around 1.5. Regarding prices, the optimal price is 11.54. Firms in the model have a mean current price of 11.75, higher than the optimal price by a small number; the standard deviation of current prices is around 0.5.

Before mergers actually happen, after a learning period, optimal prices and quantities are achieved by firms. Naturally, the adjustment process depends on the learning parameters. For example, if $\epsilon = 2$, adjustment would happen around the 250*th*

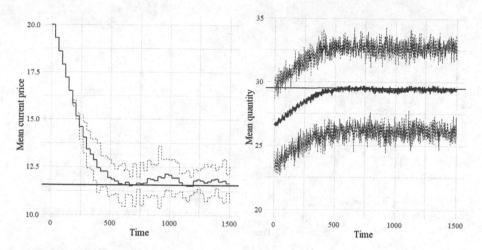

Fig. 2. Mean price (left panel) and mean quantity (right panel) as a function of time. Dashed lines represent the same time series with two standard deviations added. Black horizontal lines represent optimal values.

period. However, the standard deviations of prices and quantities would be higher. Consequently, an adjustment to the the stability parameter (θ) would be necessary to consider the higher standard deviation[6].

After Mergers

The next experiment is related to the effect of mergers on prices. Once again, 10 runs with the given initial parameters were considered, but now mergers happen only starting at the 1500*th* period and the stability condition is achieved[7]. Immediate results are given in Fig. 3, which shows prices after a merger has taken place and involves two of the three companies. Increases in prices are present for both groups that are in the industry: the merged party and the non-merged. Following the analytical model, mean prices are around 16.5 with standard deviation of 1.08 for the merging party. In the case of the non-merging firm, mean price is 13.6 and its standard deviation is 0.724.

Because this difference in prices could be due to randomness, statistical tests were conducted to help decide if the difference in means are significant. The result from a one-way ANOVA[8] test suggests that the null hypothesis of equal price means between merged and non-merged parties can be rejected with a confidence level of 95%. When considering a non-parametric test, such as Kruskal-Wallis[9], the result is the same: we reject the null hypothesis that means are equal. This suggests the changes observed after a merger are not only by chance.

---

[6] This is relevant only when mergers are allowed.
[7] This means that firms merge around the 1590th period.
[8] See [4].
[9] See [7].

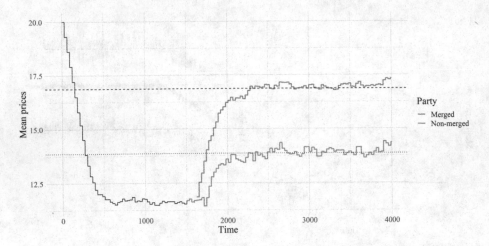

Fig. 3. Mean prices after a merger. Dashed lines represent optimal prices for each party after mergers.

As suggested by the analytical model, a merger leads to an increase in prices for both parties in the market (the merged and the non-merged). If prices are increasing, consumer's and producer's surpluses are both affected. The left panel of Fig. 4 shows the evolution of surplus for both parties. Not only consumer surplus decreases after a merger, the total welfare, which is the sum of the surpluses for both parties, decreases as shown in the right panel. This is another expected result in the analytical model.

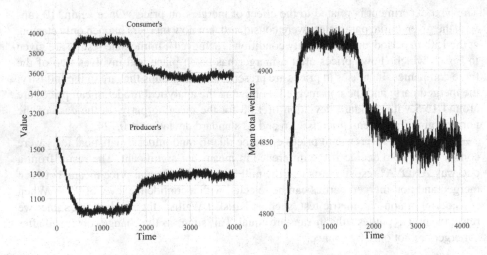

Fig. 4. Welfare measures time progression. The left panel shows the consumer's and welfare's surpluses, while the right panel shows total welfare.

4.1   Results from Cournot Competition with Differentiated Products

This simulation experiment considered an agentized version of the Cournot competion with differentiated products model. In this case, firms use the *Probe and Adjust* algorithm to define their quantities. Firms are initially engaged in competition and try to learn the optimal quantities, that is, quantities that maximize their profits. After the 1590*th* period, firms in the system merge and adjust their quantities considering the new arrangement.

Figure 5 shows the mean quantities chosen by firms in the system as a function of time. Initially, the mean quantities chosen by firms are higher than the Nash equilibrium, considering the chosen parameters. For that reason, quantities are decreasing until they're stable, which happens around the 600*th* period. When firms merge, the multi-product firms decrease their quantities. In response, if $\gamma = 10$, the firm that is outside the merger increases its sold quantity. When $\gamma = 2$, quantities for the merged firm decrease, but the firm that is not part of the merger does not increase its quantity as much. This is the predicted result in the analytical model, which attests the precision of the agent-based version, even when considering that firms are rationally bounded.

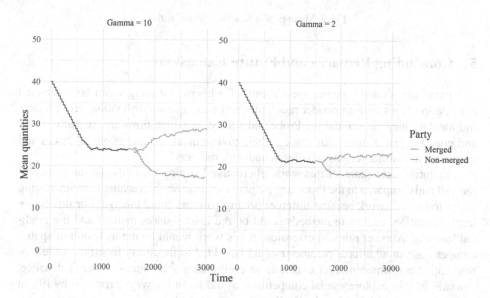

Fig. 5. Mean quantities as a function of time

The smaller the value of γ, the higher the prices after a merger. In the case that γ is sufficiently small, the profits of the merging party are increasing. An interesting pattern emerges when $\gamma = 10$: initially, the multi-product firm profit is increasing as time progresses, but it starts to decrease because the firm that is outside of the merger starts increasing its produced quantity. This is happening because the firm that is outside of the merger does not know that increasing quantities will lead to an increase in its profit. Finally, it's noticeable that mergers are unambiguously beneficial to the firm that is outside of the merger in the Cournot case. These results are observed in Fig. 6.

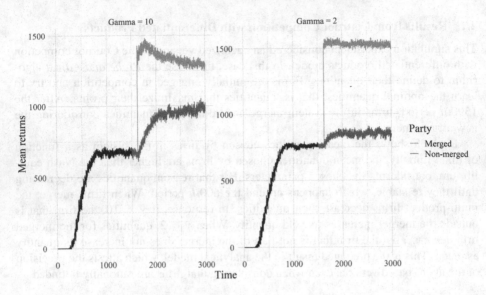

Fig. 6. Mean profits as a function of time

5   Concluding Remarks and Future Extensions

Our paper has shown how the agentization of a horizontal merger can be conducted considering firms with imperfect rationality and incomplete information. Even with a rudimentary method, such as the Probe and Adjust, in which firms are basically trying and guessing prices, firms can adapt to their environment and learn optimal prices and quantities in distinct scenarios: without and with mergers.

For future extensions of this work, alternative learning methods should be implemented and compared to the Probe and Adjust. Heterogeneous learning is an interesting extension to this work because different pricing patterns could emerge. For this future implementation, candidate methods could be the least squares method and the gradient learning. Another relevant extension to this work would be the inclusion of spatial competition: an additional parameter could be added to the utility function of the representative consumer in order to denote its preferences according to a firm's distance. Another way to explore spatial competition would be in the ways proposed by [9], in which consumers are uniformly distributed in a unit circle.

References

1. Berry, S., Levinsohn, J., Pakes, A.: Automobile prices in market equilibrium. Econometrica (1995)
2. DeSouza, S.A.: Antitrust mixed logit model. Série Estudos Econômicos (04), 4 (2009)
3. Epstein, R.J., Rubinfeld, D.: Merger simulation: a simplified approach with new applications. Antitrust Law J. **69**, 883–919 (2002)
4. Everitt, B., Hothorn, T.: A Handbook of Statistical Analyses Using R. CRC Press, Boca Raton (2009)

5. Froeb, L., Werden, G.: The effects of mergers in differentiated products industries: logit demand and merger policy. J. Law Econ. Organ. **10**, 407–26 (1994). https://doi.org/10.1093/oxfordjournals.jleo.a036857
6. Guerrero, O.A., Axtell, R.L.: Using agentization for exploring firm and labor dynamics. In: Osinga, S., Hofstede, G., Verwaart, T. (eds.) Emergent Results of Artificial Economics. Lecture Notes in Economics and Mathematical Systems, vol. 652, pp. 139–150. Springer, Berlin, Heidelberg (2011). https://doi.org/10.1007/978-3-642-21108-9_12
7. Hollander, M., Wolfe, D.: Nonparametric Statistical Methods. Wiley, New York (1973)
8. Kimbrough, S.O.: Agents, Games, and Evolution: Strategies at Work and Play. CRC Press, Boca Raton (2019)
9. Levy, D.T., Reitzes, J.D.: Anticompetitive effects of mergers in markets with localized competition. J. Law Econ. Organ. **8**(2), 427–40 (1992). https://EconPapers.repec.org/RePEc:oup:jleorg:v:8:y:1992:i:2:p:427-40
10. Motta, M.: Competition Policy: Theory and Practice. Cambridge University Press, Cambridge (2004)
11. Nardone, Claudia: Agent-based computational economics and industrial organization theory. In: Cecconi, Federico, Campennì, Marco (eds.) Information and Communication Technologies (ICT) in Economic Modeling. CSS, pp. 3–14. Springer, Cham (2019). https://doi.org/10.1007/978-3-030-22605-3_1
12. Press, W.H., Teukolsky, S.A., Vetterling, W.T., Flannery, B.P.: Numerical Recipes 3rd Edition: The Art of Scientific Computing, 3rd edn. Cambridge University Press, USA (2007)
13. Railsback, S.F., Grimm, V.: Agent-Based and Individual-Based Modeling. Princeton University Press (2019). https://www.ebook.de/de/product/34243950/steven_f_railsback_volker_grimm_agent_based_and_individual_based_modeling.html
14. Tesfatsion, L.: Chapter 16 agent-based computational economics: a constructive approach to economic theory. Handb. Comput. Econ. **2**, 831–880 (2006). https://doi.org/10.1016/S1574-0021(05)02016-2
15. Wilensky, U.: Netlogo (1999)

Multi-agent Traffic Signal Control via Distributed RL with Spatial and Temporal Feature Extraction

Yifeng Zhang, Mehul Damani, and Guillaume Sartoretti[✉]

Department of Mechanical Engineering, National University of Singapore,
Singapore, Singapore
e0576068@u.nus.edu , mpegas@nus.edu.sg
http://marmotlab.org

1 Introduction

The aim of traffic signal control (TSC) is to optimize vehicle traffic in urban road networks, via the control of traffic lights at intersections. Efficient traffic signal control can significantly reduce the detrimental impacts of traffic congestion, such as environmental pollution, passenger frustration and economic losses due to wasted time (e.g., surrounding delivery or emergency vehicles). At present, fixed-time controllers, which use offline data to fix the duration of traffic signal phases, remain the most widespread. However, urban traffic exhibits complex spatio-temporal patterns, such as peak congestion during the start and end of a workday. Fixed-time controllers [8,10] are unable to account for such dynamic patterns and as a result, there has been a recent push for adaptive TSC methods.

Reinforcement learning (RL) is one such adaptive and versatile data-driven method which has shown great promise in general robotics control. Recent works that have applied reinforcement learning to the traffic signal problem have shown great promise in alleviating congestion in a single traffic intersection [5–7,9,11]. A traffic network is composed of multiple such intersections and a network optimization problem can be broadly formulated as a centralized (single-agent) or decentralized (multi-agent) RL problem. In the centralized RL formulation, a global agent controls the entire traffic network and tries to minimize a global objective such as average trip time. However, centralized solutions for adaptive TSC are infeasible in practice due to the exponentially growing joint action and state space and the high latency associated with information centralization. To avoid the curse of dimensionality , decentralized approaches frame traffic signal control as a multi-agent RL (MARL) problem, where each agent controls a single intersection, based on locally-sensed real-time traffic conditions and communication with neighboring intersections [1,14]. In this work, we propose a framework for fully decentralized multi-agent TSC (MATSC) based on distributed reinforcement learning with parameter sharing [4], for improved scalability and

Supported by Singapore Technologies Engineering Ltd, under work package 3 of the "Urban Traffic Flow Smoothening Models" NUS-STE joint laboratory.

© Springer Nature Switzerland AG 2022
F. S. Melo and F. Fang (Eds.): AAMAS 2022 Workshops, LNAI 13441, pp. 106–113, 2022.
https://doi.org/10.1007/978-3-031-20179-0_7

performance. We design a spatial and temporal neural network, by relying on an attention mechanism and a recurrent unit, to extract spatial and temporal features about local traffic conditions at each intersection. We compare our framework with state-of-the-art MATSC methods in simulation, and show that our approach results in decreased average queue lengths and trip times, as well as increased average vehicle speeds and trip completion rates, both overall and during peak periods.

2   Method

2.1   Problem Formulation

We use a decentralized MARL formulation for MATSC. Each traffic intersection is controlled by a RL agent which only has access to local traffic conditions i.e., the agents have partial observability.

More formally, we consider the multi-agent extension of an MDP, which is characterized by a set of states, S, action sets for each of N agents, $A_1, ..., A_N$, a state transition function, $P : S \times A_1 \times ... \times A_N \rightarrow S'$, which defines the probability distribution over possible next states, given the current state and actions for each agent, and a reward function for each agent that also depends on the global state and actions of all agents, $R_i : S \times A_1 \times ... \times A_N \rightarrow R$.

We consider a partially observable variant in which an agent, i, can observe part of the system state $s_i \in \mathbf{S}$ as its observation $o_i \in \mathbf{O}$. The *state* of junction i at time step t comprises two vectors: The first one is a one-hot vector representing the current traffic phase, and the second indicates the number of vehicles on each incoming lane. In line with recent works [2], the observation space of each agent is composed of the state of its assigned junction, as well as the state of all directly connected neighboring intersections.

The action space $\mathbf{A}$ is defined as a set of non-conflicting phases. Specifically, at time step t, agent i will choose an action a_i^t from its own action space A_i as a decision for the next Δt_g period of time i.e., the intersection will be in the chosen phase from time step t to time step $t + \Delta t_g$. After the fixed duration of a given phase has elapsed, the agent may choose to continue with the same phase or choose a different phase and incur a transitional yellow phase penalty of duration Δt_y. In this work, we set Δt_g and Δt_y to $5s$ and $2s$ respectively.

Following [2], we define the reward structure as a short term metric which is calculated as the sum of the number of halting vehicles on the lane-area detectors.

2.2   Spatial and Temporal Perception Network

Given the dynamics of a traffic network, an agent (junction) needs to have both spatial and temporal awareness to make informed decisions. To enable this, we propose a network that comprises two units, a message aggregation unit for spatial feature extraction, and an RNN-based memory unit for temporal awareness. The detailed network structure is shown in Fig. 1. The attention-based message aggregation unit [13] allows the agent to learn to assign higher weights to

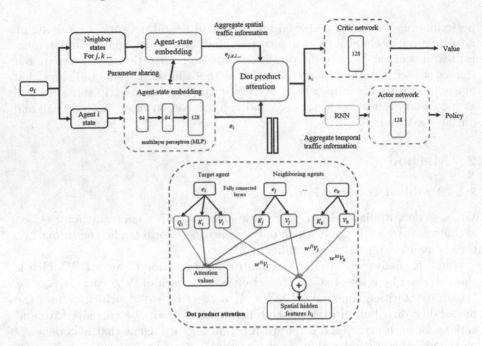

Fig. 1. Structure of the spatial-temporal perception neural network used in this work.

essential parts of the observations, i.e., concentrate more on the traffic states of neighboring intersections that might cause significant impacts on itself in future steps, while the recurrent unit allows it to utilize historical information to inform its current decision-making.

Specifically, we first map all the raw observations to a higher dimensional feature space i.e., mapping from low dimensional o_i to high dimensional e_i through multiple fully connected layers. Then, we obtain the query, key, value vectors of the attention mechanism (with same dimensions), by using three different sets of learned weights: $q_i = W_i \cdot e_i, k_i = W_k \cdot e_i, v_i = W_v \cdot e_i$. In our work, the parameters of the key, query, and value layers are shared among agents.

Next, we calculate the compatibility u_{ij} between the query q_i and k_j based on scaled dot production mechanism: $u_{ij} = (q_i^T \cdot k_j)/\sqrt{d}$, where d is the dimension of the vectors used for normalization. The attention weights for each query-key pair can be computed by: $\alpha_{ij}^h = softmax\,(u_{ij})$.

Finally, we calculate the output vector as the weighted sum of all the value vectors, using these learned attention weights: $h_i = \sum_{j \in \mathcal{N}_i} \alpha_{ij} \cdot v_j$.

Second, we rely on a recurrent neural network (here, a Gated Recurrent Unit, GRU) [3] to extract temporal features from the agents' observations. Overall, through the proposed network structure, we are able to extract features in both the spatial (using attention) and temporal (using GRU) dimensions.

2.3 Learning Framework

We use the popular PPO algorithm for training the policy [12]. PPO's update rule prevents large changes to the policy, which is particularly desirable in our distributed, parameter-sharing setting where there is significant noise in computed gradients. We use the Adam Optimizer with learning rate $5e-5$, an episode length of 720 and a discount factor(γ) of 0.95.

Inspired by some of our previous works [4], we developed a hierarchical distributed learning framework to make use of parallelization and parameter sharing. Instead of learning a separate policy for each intersection, we use parameter sharing between intersections to learn a single, universal policy common to all agents (junctions) in the network. Our distributed framework instantiates multiple low-level "workers" (meta-agents) and a high-level coordinator called the "driver". Each worker is regarded as a multi-agent system and works in an identical but independent environment. The goal of the worker is to collect the experience of all learning agents in an environment. The driver uses the shared experience of all workers to update a global shared network at the end of each episode.

In addition to significant gains in wall-clock training time, our distributed learning framework has two main advantages. First, parameter sharing between agents reduces the instability of distributed MARL associated with independent learning by preventing drastic updates to the global network and thus ensuring that the environment is relatively stable from any single agent's perspective. In addition, the (shared) network weights are updated at the end of each episode to improve the individual rewards of each agents, implicitly leading to a common policy that aims at optimizing their common decisions, thus encouraging the formation of cooperative behaviors.

Second, the structure of the distributed framework is modular and can easily be adapted to run multiple state-of-the-art RL algorithms such as A3C and SAC.

Third, we note that our learning framework aims at offline (centralized) training before online (decentralized) execution. That is, our learning agents will first be trained in simulation. Then, the trained policy can be frozen and deployed in the real world in a fully decentralized manner, i.e., based on local sensing and communication among neighboring agents. The advantages of this offline, centralized training, decentralized execution design choice are:

1. Offline training is cheap, since we do not need to be concerned about the potential threats (e.g., congestion, accidents) caused by unreasonable behaviors that would result from agents freely exploring their state-action space during early training.
2. Offline training can still allow a sim-to-real solution with high portability: the policy can be trained in simulation by relying on real-world data, if available; alternatively, the trained policy may be fine-tuned under real-world traffic conditions after deployment, to truly reach a near-optimal controller.

3   Experiments and Discussion

We conducted our simulation experiments using a Manhattan traffic network based on the benchmark method MA2C, and the same exact traffic demand [2].

As illustrated in Fig. 2, there are a total of 25 homogeneous signalized intersections in this grid network, where each one is formed by two-laned, horizontal arterial streets with a speed limit of 20 m/s, and one-laned, vertical avenues with a speed limit of 11 m/s. Each intersection contains five permissible phases: East-West straight phase, East-West left-turn phase, and three straight and left-turn phases for East, West, and North-South, respectively. The lane-area detectors are installed near the stop line of the intersection, with a length of 50 m, which are shown as the blue rectangles in Fig. 2.

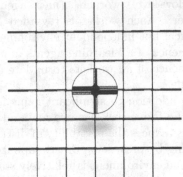

We compare our method (dDRL-Att) with a conventional greedy controller (one-step optimal controller with respect to the same metric used by our dDRL approach) and three learning based methods (MA2C, IA2C, and IQL-LR) [2]. These last three baselines are decentralized multi-agent RL algorithms, where each agents learns an independent policy depending on its local traffic conditions. IQL-LR (linear regression based independent Q-learning) is a fully scalable Q-learning algorithm, where each local agent

Fig. 2. Simulated grid traffic scenario with 25 homogeneous intersections, adapted from the benchmarks used in [2].

learns a local Q network based on its own action, while regarding other agents as part of the environment. IA2C extends the idea of IQL to the advantage actor-critic (A2C) algorithm, where the actor network directly learns a policy that maps the input state to a probability distribution over actions, while the critic network learns to predict the state value as a baseline for calculating the advantage value guiding the actor's training. However, IQL-LR and IA2C suffer from non-stationarity, which mainly stems from partial observability and limited communication among agents. Multi-agent A2C (MA2C) attempts to mitigate this common problem of independent learning algorithms by proposing two improvements. Firstly, MA2C includes the states and the sampled latest policies (fingerprints) of neighboring agents in the local state of each agent. Second, it introduces a spatial discount factor to decrease the impacts of both the observations and rewards of neighboring agents, thus encouraging agents to concentrate more on their local traffic conditions. In some ways, we note that the attention weights learned by our approach play a similar role as this spatial factor, but with the added advantage that our weights are dynamic, and thus can adapt to the different states of neighboring agents.

We also include a non-attention version of our method (dDRL) for a simple ablation study on this aspect. The comparison test results are shown in Fig. 3,

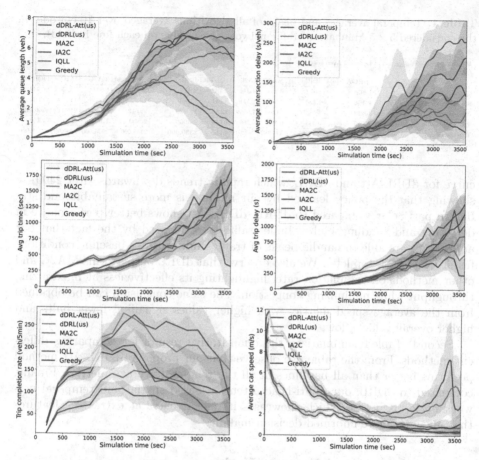

Fig. 3. Evaluation results for 10 episodes in the considered 5 × 5 Manhattan network. The solid lines show average values, while the standard deviations are shaded.

where we measure and record the traffic metrics at each simulation step and then calculate the averages and deviations over a fixed set of 10 test episodes (i.e., same traffic conditions for all algorithms for fair comparison).

From these evaluation results, we first observe that the queue length and delay time curves of both IQL-LR and IA2C show a monotonically increasing trend. This trend indicates that as the simulation time increases, the vehicles gradually start accumulating on the incoming lanes, leading to growing conges-tion. Compared to IQL-LR and IA2C, the queue length and delay time curves for MA2C rise significantly slower, indicating that MA2C is able to learn an effective policy. However, it still lacks the ability to reduce congestion in sat-urated networks, especially when the traffic gets heavier after 2000s.s. Finally, our distributed deep RL methods (dDRL-Att and dDRL) outperform the above-discussed baselines on handling peak-time traffic as well as recovering from sat-urated and congested traffic networks. This is evident from the queue length

Table 1. Temporal averages and peaks of all algorithms averaged over 10 episodes in the considered 5×5 Manhattan traffic network (best value on each line in bold).

Metrics (Average)	Temporal Averages					Temporal Peaks				
	dDRL-Att (us)	dDRL (us)	MA2C	IA2C	IQL-LR	dDRL-Att (us)	dDRL (us)	MA2C	IA2C	IQL-LR
Queue length (veh)	**1.800**	1.974	3.204	4.214	3.762	**3.816**	3.939	5.944	7.555	6.942
Speed (m/s)	**4.468**	4.223	2.003	1.571	3.041	13.361	**14.176**	11.656	11.513	13.978
Intersection delay (sec)	**30.984**	38.837	34.985	72.959	92.591	**70.929**	98.612	124.005	216.112	266.912
Trip completion rate (veh/s)	**0.990**	0.955	0.546	0.366	0.634	**2.300**	**2.300**	1.700	1.400	2.100
Trip time (sec)	**447.519**	460.735	727.387	788.501	462.862	**2330.000**	2745.000	2938.000	3445.000	2373.000

curve for dDRL-Att and dDRL, which rapidly trends downwards after 2400 s,s, showing that the policy learned by our method is more sustainable and stable. In particular, compared to dDRL, dDRL-Att shows better performance in detecting and handling high-volume traffic, as evidenced by the fact that our approaches are able to handle peaks of traffic and return to baseline conditions faster (see Fig. 3, top left). We also observe that dDRL outperforms MA2C and other methods on completion rate, highlighting its effectiveness at maximizing throughput while minimizing congestion. The same conclusion can be obtained from the average speed metric, with higher values signalling fewer halts and higher overall vehicle flow.

Second, Table 1 summarizes the quantitative results for comparing different methods. From the table, we observe that our proposed dDRL-Att method performs better than all baselines in all the temporal average metrics. Finally, compared to dDRL, our multi-head-attention-based spatial and temporal network performs better, which showcases its effectiveness at extracting features that allow for more informed decision making.

4   Conclusion and Future Work

In this work, we introduced a framework for fully decentralized multi-agent TSC (MATSC) based on distributed reinforcement learning with parameter sharing, which relies on an attention mechanism and a recurrent unit for the extraction of spatial and temporal features. Through experimental results, we demonstrated that our proposed framework results in decreased average queue lengths, and increased average vehicle speeds and trip completion rates. Future work will focus on developing and improving techniques such as credit assignment that allow for increased cooperative behavior between agents in an effort to achieve better performance on global long-term metrics while staying in the regime of partial observability and local sensing.

Acknowledgements. Authors would like to thank Dr. Teng Teck Hou for his feedback on earlier drafts of this manuscript as well as for research discussions during the elaboration of this work.

References

1. Camponogara, E., Kraus, W.: Distributed learning agents in urban traffic control. In: Pires, F.M., Abreu, S. (eds.) EPIA 2003. LNCS (LNAI), vol. 2902, pp. 324–335. Springer, Heidelberg (2003). https://doi.org/10.1007/978-3-540-24580-3_38
2. Chu, T., Wang, J., Codecà, L., Li, Z.: Multi-agent deep reinforcement learning for large-scale traffic signal control. IEEE Trans. Intell. Transp. Syst. **21**(3), 1086–1095 (2019)
3. Chung, J., Gulcehre, C., Cho, K., Bengio, Y.: Empirical evaluation of gated recurrent neural networks on sequence modeling. arXiv preprint arXiv:1412.3555 (2014)
4. Damani, M., Luo, Z., Wenzel, E., Sartoretti, G.: PRIMAL2: pathfinding via Reinforcement and Imitation Multi-Agent Learning - Lifelong. IEEE Robot. Autom. Lett. **6**(2), 2666–2673 (2021). https://doi.org/10.1109/LRA.2021.3062803
5. Gao, J., Shen, Y., Liu, J., Ito, M., Shiratori, N.: Adaptive traffic signal control: deep reinforcement learning algorithm with experience replay and target network. arXiv preprint arXiv:1705.02755 (2017)
6. Garg, D., Chli, M., Vogiatzis, G.: Deep reinforcement learning for autonomous traffic light control. In: 2018 3rd IEEE International Conference on Intelligent Transportation Engineering (ICITE), pp. 214–218. IEEE (2018)
7. Genders, W., Razavi, S.: Evaluating reinforcement learning state representations for adaptive traffic signal control. Procedia Comput. Sci. **130**, 26–33 (2018)
8. Hunt, P., Robertson, D., Bretherton, R., Winton, R.: Scoot-a traffic responsive method of coordinating signals. Technical report (1981)
9. Li, L., Lv, Y., Wang, F.Y.: Traffic signal timing via deep reinforcement learning. IEEE/CAA J. Automatica Sinica **3**(3), 247–254 (2016)
10. Luk, J.: Two traffic-responsive area traffic control methods: scat and scoot. Traffic Eng. Control **25**(1) (1984)
11. Mousavi, S.S., Schukat, M., Howley, E.: Traffic light control using deep policy-gradient and value-function-based reinforcement learning. IET Intel. Transp. Syst. **11**(7), 417–423 (2017)
12. Schulman, J., Wolski, F., Dhariwal, P., Radford, A., Klimov, O.: Proximal policy optimization algorithms. arXiv preprint arXiv:1707.06347 (2017)
13. Vaswani, A., et al.: Attention is all you need. Adv. Neural Inf. Process. Syst. **30** (2017)
14. Wei, H., et al.: Presslight: learning max pressure control to coordinate traffic signals in arterial network. In: Proceedings of the 25th ACM SIGKDD International Conference on Knowledge Discovery & Data Mining, pp. 1290–1298 (2019)

About Digital Twins, Agents, and Multiagent Systems: A Cross-Fertilisation Journey

Stefano Mariani[1]([✉])[iD], Marco Picone[1][iD], and Alessandro Ricci[2][iD]

[1] Department of Sciences and Methods of Engineering, University of Modena and Reggio Emilia, Modena, Italy
{stefano.mariani,marco.picone}@unimore.it
[2] Department of Computer Science and Engineering, University of Bologna, Bologna, Italy
a.ricci@unibo.it

Abstract. Digital Twins (DTs) are emerging as a fundamental brick of engineering cyber-physical systems, but their notion is still mostly bound to specific business domains (e.g. manufacturing), goals (e.g. product design), or applications (e.g. the Internet of Things). As such, their value as general purpose engineering abstractions is yet to be fully revealed. In this paper, we relate DTs with agents and multiagent systems, as the latter are arguably the most rich abstractions available for the engineering of complex socio-technical and cyber-physical systems, and the former could both fill in some gaps in agent-oriented engineering and benefit from an agent-oriented interpretation—in a cross-fertilisation journey.

Keywords: Digital twin · Agent · Multiagent system · Cyber-physical system

1 Introduction

In the last decade, the Digital Twin (DT) paradigm has been explored in different domains [23,35,39] as an approach to *virtualise* entities existing in the real world – creating software counterparts meant to be *faithful* digital replicas, deeply *intertwined* with their physical twin [15,16,26] –, and recently it has been better shaped through a well-defined set of abstract capabilities and responsibilities for DTs. The agent-oriented engineering (AOE) paradigm is instead well known since at least three decades, and is nowadays a reference for the engineering of complex socio-technical and cyber-physical systems [5,19,44].

Intelligent agents and multiagent systems (MASs) [19] can exploit DTs as a *virtual environment* (or, application environment [42]) enabling access and interaction with the physical world. In this view, a DT functions first of all as a *shared medium* used by agents to observe and act upon the Physical Assets (PAs) structuring the physical world. Besides, a DT may provide further higher-level

This work has been partially supported by the MIUR PRIN 2017 Project N. 2017KRC7KT "Fluidware".

© Springer Nature Switzerland AG 2022
F. S. Melo and F. Fang (Eds.): AAMAS 2022 Workshops, LNAI 13441, pp. 114–129, 2022.
https://doi.org/10.1007/978-3-031-20179-0_8

functionalities with respect to the associated PA – conceptually *augmenting* its native capabilities – that could be exploited by agents to support their reasoning and decision making upon the resulting *cyber-physical system* [2].

However, it is also possible to envision the opposite scenario: DTs exploiting agents and MASs to deliver more intelligent functionalities, as a way of realising the vision of *cognitive* DTs [1,11], that refers to those DTs that *autonomously* perform some intelligent tasks within the context of the PA—related to e.g. smart management, maintenance, and optimisation of performances. Even DTs modelled or implemented as agents have been reported in the literature [3,36].

Whatever the case, we argue that the agent and DT abstractions lend themselves to a clear *separation of concerns* from a design perspective, depicted in Fig. 1 —that we develop in this paper, especially in Sects. 3.1 and 4.1: DTs operate within the boundaries set by the *local context* of their associated PA, for instance in terms of which information they can access and which actions they can carry out, that they are perfectly aware of due to their deep bond with the PA. Agents, instead, do not have such limitation: for instance, they may access information provided by other agents or third party services, as well as request others to carry out actions on their behalf. However, agents do not have the knowledge about the cyber-physical context as DTs do. This is the main motivation for their *synergistic* exploitation—as well as for the discussion put forward in this paper.

Accordingly, in this perspective paper we highlight the importance to identify responsibility and operational boundaries between DTs, and agents and MAS – briefly described in Sect. 2 –, and shed light on their existing and potential synergies by analysing both perspectives of what DTs can do for agents and MASs (Sect. 3), and what agents and MASs can do for DTs (Sect. 4). Then, we speculate about more exotic research lines that are currently under-explored,

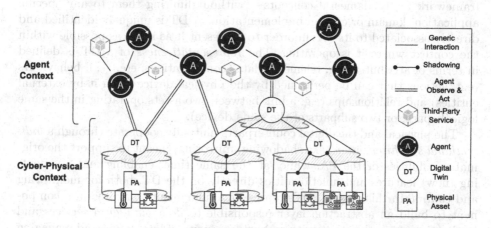

Fig. 1. *Separation of concerns*: DTs operate within the boundaries set by the *local context* of the associated PA, whereas agents operate within the boundaries of the *global context* of the whole application set by the application designer.

but could prove to be meaningful (Sect. 5). Finally, we conclude the paper with some final remarks (Sect. 6).

We emphasise that the upcoming figures do not depict a system *architecture* (not even an abstract one), but are a graphical way to represent the *mindset* that the system designer should keep in mind when adopting the perspective described in the corresponding Section. In fact, depending on specific deployment constraints and implementation requirements, each perspective could give raise to different architectures (e.g. deployment at the Edge vs. on Cloud). This aspect is better discussed in Sect. 6.

2 Background

DTs are well known outside of the MAS community, where they started to gain traction much more recently. Here we provide a brief account of both traditional DT literature and agent-oriented exploitation of DTs.

2.1 Digital Twins Outside of MAS

The concept of Digital Twin (DT), introduced between 1999 and 2002 [38], has been recently revisited due to the advent of the Internet of Things (IoT) and the quick migration to a technological ecosystem where the effective collaboration between *cyber and physical layers* represent a fundamental enabler for the next generation of applications.

A DT represents the *digitised software replica* of a Physical Asset (PA) with the responsibility to clone available resources and functionalities, and to extend existing behaviours with new capabilities. DTs have been recently characterised in the literature [25,26,34] with a set of responsibilities and capabilities, with the aim to identify a common set of features and to provide a unified conceptual framework for fundamental concepts— without limiting them to any specific application domain or custom implementation. A DT is uniquely identified and directly associated to its PA, in order to represent it as much as possible within the context where it is operating. The *representativeness* of a DT is defined in terms of attributes (e.g. telemetry data, configurations, etc. ...), behaviours (e.g. actions that can be performed by the physical device or on it by external entities) and relationships (e.g. a link between two assets operating in the same logical space, or two subparts of the same device).

The physical and the digital counterparts mutually cooperate through a *bidirectional synchronization* (aka shadowing, mirroring) meant to support the original capabilities of the PA, while, at the same time, enabling and augmenting (new) features and functionalities directly on the DT, both for monitoring and *control*. In this context, DTs represent a fundamental architectural component to build an abstraction layer responsible to *decouple* digital services and applications from the complexity and heterogeneity of interacting and managing deployed PAs. They allow observers and connected services to easily integrate cyber-physical behaviours in their application logic, and to design and execute high level functionalities with no concern for the complexity of end devices.

2.2 Digital Twins Within MAS

A good overview of DTs exploitation in MAS to date is given in [20], although focussed on the manufacturing domain. There, DTs are mostly assumed to always undergo a process of "agentification" meant to improve DTs capabilities, e.g. inherit agents' abilities to negotiate and interact with other peers of the same system. On the contrary, the Activity-resource-type-instance (ARTI) architecture [40] starts to foster synergy of DTs with MAS as distinct entities, by differentiating among *decision-making* "agents" from *reality-reflection* "beings" (the DTs), a distinction similar to the separation of concerns we described in the introduction. Also reference [13] promotes the idea that through a MAS a set of DTs can create a network named as "asset fleet", essentially enabling DTs to obtain information about events that have not affected them yet, as a way to improve their individual knowledge of the environment. Another review in favour of a synergistic exploitation of MAS and DTs, while recognising the need for further research along this line, argues that agents and MAS are good examples of how autonomous decision-making can be modelled and implemented based on digital representations of physical entities [18]—as DTs are.

Another literature review [29], explicitly targeting the supply chain business domain, sums up well how MAS and DTs are currently mostly exploited in synergy (emphasis added)—also outside of the supply chain domain:

> "Since supply chains are now building with increasingly *complex and collaborative interdependencies*, Agent-Based Models are an extremely useful tool when representing such relationships [...]. While Digital Twins are new solutions elements for enable real-time digital monitoring and control or an automatic decision maker with a higher efficiency and accuracy".

The literature also accounts for works that apply agents for modelling, designing, implementing, or exploiting DTs. In [3] BDI agents – being BDI (Belief-Desire-Intention) a main model/architecture adopted to implement knowledge-based rational agents [31] – are proposed to represent DTs of real-life organisations, claiming that beliefs, desires, and intentions are suitable abstractions for characterising mental attitudes of anthropomorphic organisations. A similar approach is proposed in [36] where agents are adopted as a metaphor to revise the structure of a DT in an autonomous, behaviour-centred perspective encapsulating the inherent agent's perception–decision–action cycle and intelligence. Finally, a previous work of co-authors [10] builds agent-based DTs for the healthcare domain. In [9] instead, a whole MAS is used to implement the DT of a whole city transportation system. A similar approach is taken in reference [41] to realise the concept of "communicating material" in the energy supply business domain.

3   DTs for Agents and Multiagent Systems

The most natural way to relate DTs with agents and MASs is via the *environment* abstraction of a MAS, as depicted in Fig. 2: there, DTs *encapsulate* PAs

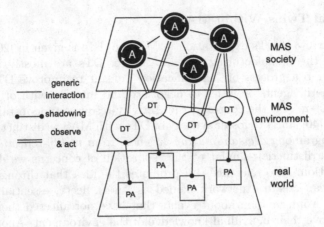

Fig. 2. DTs as MAS *environment*: they encapsulate cyber-physical resources and mediate access to them.

state (properties, relationships) and behaviour, and *decouple* agents access to PAs, both for monitoring and control. Under this perspective, DTs work as the software engineering abstraction enabling *cyber-physical modelling* of the MAS environment and supporting agents interaction with it.

3.1 Individual Perspective

In this sense, DTs are akin to the *artefacts* of the A&A metamodel [27], as they are used by agents to bidirectionally interact with the physical layer, augment their functionalities, or coordinate their execution according to the target goals. They can be also exploited by system designers to either give structure and dynamics to the MAS computational environment, or model and enable access to a physical environment the MAS has to cope with. However, they are also potentially more powerful than artefacts, as they are strongly coupled with their associated PA: DTs should guarantee that changes in the PA are promptly reflected in the DT, and, the other way around, that changes to the DT affect the associated PA when due. Artefacts, instead, do not have this deep bond with the physical world by design, but simply are a model of an object of interest that is not worth to be modelled as an agent—according to the system designer.

To be more practical, by accepting this view an agent-oriented application designer could naturally ascribe to agents tasks requiring abilities such as planning, reasoning and inference, complex analytics, and any other task that can be placed under the umbrella term "decision making", and to DTs functions such as monitoring, events logging, remote operations, and any other task related to perception and control of the associated PA (hence, of the *environment*), as exemplified in Fig. 3—albeit not exhaustively. However, there are also a whole bunch of tasks that are not so easy to be ascribed to either entity: *prediction* capabilities and *simulation* of alternative scenarios or courses of actions, for

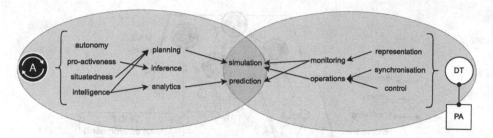

Fig. 3. Synergistic exploitation of agents and DT's capabilities: each abstraction is best suited for different tasks, that complement each other.

instance, are examples of complex functionalities that can be given to agents, by leveraging their intelligence, to DTs, by leveraging their entanglement with PAs, but also to an agent-DT ensemble, where each entity contributes with its own capabilities, and it is their *synergistic exploitation* that delivers the sought functionality optimally—as depicted in Fig. 3.

For instance, let us assume that the goal to achieve is some sort of "what-if" analysis in a generic industry 4.0 deployment: an agent may reason about which controlled variables (actuators in the physical environment) need to change to reflect the simulated scenario, then send the appropriate control commands to a DT that generates the associated effects in the digital world, *without* affecting the actual PA (it is a simulation), so as to enable observation of uncontrolled variables (sensors in the physical environment) in the alternative, simulated scenario. This kind of "on/off switch" enabling to bind/unbind the DT to its PA on a temporary basis already represents a software engineering challenge *per se*. Then, when simulation results are satisfactory, DTs may be exploited to actually bring about, on their associated PAs, the actions corresponding to the information gained from the what-if analysis—closing the feedback loop realising the idea of *actionable knowledge*. Neither component would achieve the same result alone: the agent may not known the inner dynamics of the PA, and the DT may lack the knowledge of the application domain required to understand how to generate the what-if scenario.

3.2   System Perspective

The literature on DTs is abundant and mostly settled on what to expect from an *individual* DT, but not much is said about how to structure complex shadowing scenarios besides simple aggregation of DTs: is there one DT for each PA? Can DTs be somehow "linked together" to give structure to the mirrored environment? Should such structure, if any, be hierarchical? Can it be changed dynamically and spontaneously by DTs themselves, to reflect endogenous dynamics between the associated PAs? The most domain-agnostic view of these issues is given in the *Web of Digital Twins* (WoDT) vision [32], where DTs are seen as entities interlinked in a *web of semantic, dynamic relationships*, that enable structuring a dynamic application domain.

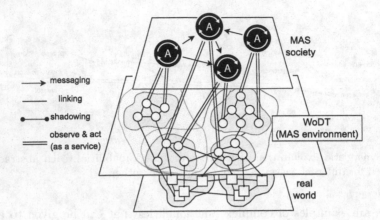

Fig. 4. *Web of Digital Twins* as MAS environment: application-dependent, semantic overlay networks are dynamically established amongst DTs depending on the cyber-physical system dynamics and application goals.

According to the WoDT vision, as depicted in Fig. 4, a layer of networked DTs work as the interface between applications (either agent-oriented or not) and the physical environment they must cope with, thus decoupling the two layers while possibly providing augmented and cross-domain functionalities. The DTs network in WoDT is a *knowledge graph* [17] – that is, a semantic network where links amongst nodes in the graph have meaning specified by an accompanying domain and application-specific ontology – created through both design-time relationships reflecting the structure of the PAs in the physical environment, and run-time linking operations spontaneously carried out by DTs (or requested by applications) to timely reflect the ever evolving environment dynamics. The result of this semantic linking is the dynamic creation of semantic overlay networks that applications can navigate to makes sense of the physical world and affect it according to their goals, while exploiting the added functionalities provided by DTs, and most importantly disregarding any specific heterogeneity and technicalities of interactions with the associated PAs. For instance, a WoDT could be deployed to provide basic services in the context of a smart city, such as opportunistic ride-sharing, smart parking, intelligent intersection management, and the like. There, DTs would be created for vehicles, Road Side Units (RSUs), and possibly people, and the links amongst some of these DTs would only be established dynamically, depending on what happens in the physical world. As an example, the DT of an intersection may create a link with all incoming vehicles as soon as they are detected via monitoring cameras, with the semantics that such vehicles need orchestration of that intersection to cross safely.

Remarks. This interpretation where DTs work as the environment abstraction in a MAS is not the only possible one, but the most natural from the standpoint of agent-oriented engineering (AOE). Under this perspective, DTs bring to MAS a powerful engineering abstraction on top of which to design interaction with a physical environment, both as regards observation to gather information and plan actions accordingly, and control given the agents' goals. In Sect. 5 further perspectives are discussed.

4 Agents and Multiagent Systems for DTs

When switching to a DT-oriented perspective (opposite to the AOE one adopted in previous Section), the most natural way to exploit agents for the benefit of DTs is possibly as *enablers of intelligent behaviours* and as *orchestrators and mediators* of DTs interactions.

4.1 Individual Perspective

Besides providing a digital replica of a PA, always synchronised with its physical counterpart, the literature about DTs often times mention their capability to provide *intelligent* functionalities [12] and/or to *augment* the innate capabilities of the associated PA via software. Prediction of possible future events, detection of anomalies, and simulation of alternative configurations of a PA are common examples of such added capabilities. However, there is no consensus yet on a standard and application independent way of delivering such functionalities, and on a way to *encapsulate* them in reusable components available across applications and serving multiple DTs. In other words, there is not yet a shared model of how to *deliver intelligence* in the context of DTs. Sometimes it is achieved by hard-wiring machine learning training pipelines or models into DTs [21], some other time it is an external service built ad-hoc for the application at hand [23].

Agents, instead, do offer reference models/architectures for defining intelligent behaviours, such as the BDI model [31], and most importantly allow to encapsulate the required intelligence in an autonomous and independent component, that may be then requested to provide its functionalities in a loosely coupled way, *as a service* to multiple DTs concurrently. Under this perspective, as depicted in Fig. 5, agents are *peers* of DTs, offering *services* meant to be exploited to get whatever the required intelligent behaviour is, e.g., prediction of future possible events based on historical data threaded by the DT, simulation of alternative configurations based on agent's own reasoning and inference capabilities, planning of complex sequence of actions to be undertaken on the PAs associated to the DTs.

From a design perspective, the same considerations depicted in Fig. 3 apply: agents and DTs have *complementary* capabilities that the designer of the application at hand can exploit synergistically to achieve the intended goal in the best possible way, and by adequately separating concerns. However, the solution designed follows a very different paradigm from the one depicted in Fig. 2: there, applications are structured around agents, hence agents are the one responsible to achieve the application goals (possibly exploiting DTs for accessing and controlling the physical world), whereas here, instead, the application revolves around DTs and the services and functionalities they deliver, while agents are

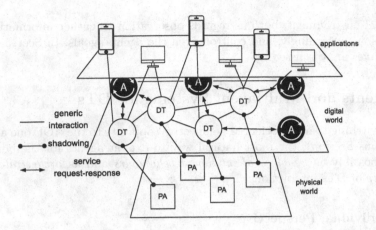

Fig. 5. Agents as enablers of DT intelligence: DTs delegate to agents complex forms of reasoning and prediction that require broader context with respect to the local one accessible to DTs.

transparent to the user (i.e. she may not even be aware that DTs are exploiting agents' capabilities to deliver their intelligent functionalities).

4.2 System Perspective

In non trivial cyber-physical systems a multitude of DTs co-exist, are possibly distributed across space, and could be created and disposed dynamically—in an open systems perspective. There, it is not always clear, according to the available literature [26], how DTs should interact to realise the application goals as well as who is responsible for their lifecycle: is there an orchestrator? Is it a DT itself? Can DTs be composed somehow akin to service composition patterns [43]? Is composition the only interaction pattern they need?

All of these open questions could find an answer in AOE. Agents can work as the *orchestrators* responsible to handle DTs lifecycle in compliance with the goals and constraints put forth by applications. Agents can also *mediate* DTs interactions, as the MAS literature is abundant in communication protocols and coordination models going beyond service composition schemes [4,6,8,28,44].

For instance, Fig. 6a depicts how a *logical* interaction between DTs (the bold dashed line) may actually unfold as a complex coordination protocol – i.e. a structured sequence of interactions – carried out by agents on DTs behalf (the greyed out area with lines and arrows). In this way, that is, by delegating administration of interactions to agents, DTs can simply express the intended interaction semantics and let agents figure out the actual communication or coordination protocol required to enact such semantics. As an example, a DT may express the need for an information along with a minimum degree of confidence that the information is truthful, and let agent carry out a ContractNet protocol [7] amongst other DTs to find the one with the highest degree of confidence.

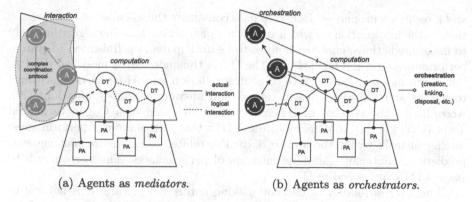

(a) Agents as *mediators*.   (b) Agents as *orchestrators*.

Fig. 6. Agents for DTs orchestration and coordination: communication and coordination protocols from MAS literature support advanced forms of DTs opportunistic and temporary aggregation, besides traditional service composition.

Figure 6b instead depicts how DTs could be orchestrated by agents, decoupling the "DevOps" logic from the application logic (the dotted lines with diamond ending). Agents may be responsible for DTs creation, linking, disposal, replication, relocation (across the Edge-to-Cloud deployment spectrum), etc. as they are aware of the application context—extending beyond the PA context that DTs are aware of.

5   Research Directions

The opportunities for synergistic exploitation of agents, MAS, and DTs presented so far are possibly the most natural to think about given the nature of agents and DTs as described by the main body of related literature. In summary, the cross-fertilisation opportunities brought to light in this paper revolve around two core ideas: on the one hand, adopt DTs as the engineering abstraction to structure and encapsulate the resources and dynamics of the cyber-phsyical system at hand, while, on the other hand, rely on agents as the engineering abstraction to encapsulate decision making towards realisation of the application goals. The contact point between the two lies in the fact that such decision making is affected by (and must affect, in turn) the cyber-physical system itself. Hence synergistic exploitation of agents and DTs is more of a requirement than a desirable design choice, for these kinds of systems.

However, this perspective is not the only one worth exploring, nor the broadest one, hence we here provide some discussions about ongoing or potentially viable research activities. regarding agents and DTs integration.

5.1   Cognitive DTs

Recent advancements in IoT, big data, and machine learning have significantly contributed to the improvements in DTs regarding their real-time capabilities and forecasting properties. Collected data constitute the so-called *digital threads*, that is, the information on which simulation or machine learning algorithms rely to make predictions, enabling failures to be anticipated, optimisation of system performance, and the like [34,39]. The DT is thus not only a model of the PA, but it can *autonomously* evolve through simulation and AI-enabled algorithms, to understand the world, learn, reason, and answer to *what-if* questions [22]. Accordingly, the concept of DT has evolved into that of a Cognitive Digital Twin (CDT) [1,11], that refers to those DTs that autonomously perform some intelligent tasks within the context of the PA, related to e.g. smart management, predictive maintenance, and optimisation of performances. This corresponds to stage 4 DTs envisioned in [34].

Whenever autonomy of decision making enters the picture, it is natural to look at agent models and technologies to deliver such autonomy. Hence, research about the possibly many architectural relationships between agents and DTs must be carried out. Embedding of agents inside DTs, service-oriented integration, hypermedia-based cooperation, and artefact-mediated coordination are some of the possibilities to let DTs gain advantage of agents autonomy, and agents to exploit DTs deep bond with the cyber-physical system composed by the mirrored PAs.

5.2   Anticipatory Planning

CDTs are strongly related to prediction and simulation capabilities. But these capabilities may enable another advanced form of reasoning: anticipatory planning, that is, planning not in reactions to present contextual conditions, such as when the beliefs of an agent change due to novel perceptions, but in anticipation to future likely events.

In fact, when DTs are endowed with the capability to predict future configurations, states, or behaviours of the mirrored PA, intelligent agents may exploit such predictions to undertake anticipatory coordination actions, meant to improve their or the system performance before disruption of the status quo actually occurs [24]. Furthermore, intelligent agents may exploit the ability of DTs to digitally replicate the PA to simulate alternative configurations, operating processes, or scenarios, with the aim to carry out what-if analyses without affecting the actual PA.

5.3   Sociotechnical Systems

DTs are commonly associated to PAs intended as physical objects of the physical reality, such as sensor and actuator devices, products, or machinery, that they *mirror* (or *shadow*) in the digital plane. But in a *socio-technical system*, where people and organisations are involved, there may be more than these kinds of physical objects to mirror. As fostered in [32], PAs are anything worth digitising according to the application at hand, there including processes, people, organisations, as well as virtual resources (e.g. a database, a virtual machine, a server).

The literature is already starting to consider this option, as in the case of the work in [10] were the DT of a patient is modelled. Other works consider whole organisations and systems, such as in the context of smart cities [35]. Also works considering DTs of (production) processes are available [30], as further witness of the increasing broadening of term "Physical Asset".

In the specific context of agent societies, the mirroring opportunities are even more: DTs may mirror communication channels or infrastructures, such as an event-bus or a messaging service. However, why mirroring such virtual resources, since they are already digital, is a question that should be answered as soon as possible if one would like to explore this research line.

5.4   Mirror Worlds

By pushing to its limits the idea to have a pervasive substrate of DTs, not only mirroring physical objects and equipment, but also providing digital artefacts embodied in some way into our physical reality (e.g. via holograms), we get to the idea of *mirror worlds*, as originally inspired by D. Gelernter in [14], and further explored and developed in the context of agents and multiagent systems in [33]. Following Gelernter, mirror worlds are *"software models of some chunk of reality"* [14], that is: *"a true-to-life mirror image trapped inside a computer"*, which can be then viewed, zoomed, analysed by real-world inhabitants with the help of proper software (autonomous) assistant agents. The primary objective of a mirror world is to strongly impact the lives of the citizens of the real world, offering them the possibility to exploit software tools and functionalities provided by the mirror world, generically, to tackle the increasing life complexity. The same vision applies to Web of Digital Twins [32], which could be considered a concrete approach to design and develop mirror worlds under this perspective.

5.5   Standardisation and Interoperability

As clearly reviewed and pointed out also in [26], the literature is conceptually aligned on an idea and on the importance of DT in multiple fields, but the definition of an interoperable set of properties, behaviours, and a standard description language is an ongoing activity still far from full realisation [37].

The fragmentation of existing solutions is mostly related to their specificity for a target sector and the missing detailed definition of how DTs should be represented and operate. On the one hand, the resulting trend generates innovative approaches in disparate fields. However, on the other hand, it limits the real potential of uniformed DTs by creating an unnecessary substrate of heterogeneous approaches. The opportunity to define an uniform and interoperable layer of DTs is a fundamental building block if we really aim to exploit them through a synergistic interaction with intelligent agents and multiagent systems. On one hand, we want to delegate to DTs the complexity of managing and interacting PAs, on the other hand we cannot force MASs to embrace the complexity of handling a plethora of heterogeneous and isolated DT platforms.

In order to obtain an effective multi-layer architecture where PAs, DTs, and MASs can seamlessly cooperate there is the tangible need to start working on existing platforms on both sides in order to identify how existing functionalities and models can be extended to work together through the use of standardized solutions and avoiding the creation of an additional substrate of custom integration modules. Within this context, standardisation of DTs may be for agent-environment interaction what FIPA has been for agent-agent interaction [4].

6   Concluding Remarks and Outlook

In this paper, we analysed the potential synergies between agents and DTs, and multiagent systems and (networks of) DTs, to reason about both the individual and collective (system) level. Our aim was to shed light on the responsibilities each abstraction has with respect to cyber-physical systems engineering, and on the motivations and expected benefits of their integration. As such integration can be realised in many different ways, as witnessed by the extremely heterogeneous literature about DTs exploitation within MAS, we tried to discuss the available alternatives starting from the most natural ones, that is, those that (seem to) best adhere to the defining characteristics of the agent and DT abstractions. Nevertheless, we also briefly commented on more exploratory research activities that need to be carried out to exhaustively carry on research about DTs and agents integration.

We did so in the attempt to clarify the mindset that system engineers should have while designing their solution, not as the proposal of a reference architecture. In fact, many are the factors that influence integration of AOE and DTs at the architectural level, hence it is more likely that each perspective described in this paper gives raise to slightly different architectures, than that each perspective has a direct mapping with one and only admissible architecture. Defining a methodology to devise out a specific architecture given a perspective and some constraints (regarding deployment, implementation, application requirements, etc.) would indeed be an interesting research thread.

Accordingly, we hope this perspective paper can stimulate critical discussion in the MAS community and to the wider audience interested in DTs-based software solutions, regarding this emerging and broadening novel characterisation of Digital Twins, that cannot be ignored.

Acknowledgements. We would like to thank the anonymous referees and the EMAS community for their insightful comments that helped to improve the manuscript.

References

1. Abburu, S., Berre, A.J., Jacoby, M., Roman, D., Stojanovic, L., Stojanovic, N.: COGNITWIN - hybrid and cognitive digital twins for the process industry. In: IEEE International Conference on Engineering, Technology and Innovation (ICE/ITMC), pp. 1–8 (2020)

2. Ahmed, S.H., Kim, G., Kim, D.: Cyber physical system: architecture, applications and research challenges. In: Proceedings of the IFIP Wireless Days, WD 2013, Valencia, Spain, 13–15 November 2013, pp. 1–5. IEEE (2013). https://doi.org/10.1109/WD.2013.6686528

3. Alelaimat, A., Ghose, A., Dam, H.K.: Abductive design of BDI agent-based digital twins of organizations. In: PRIMA 2020: Principles and Practice of Multi-Agent Systems - 23rd International Conference. LNCS, vol. 12568, pp. 377–385. Springer, Cham (2020). https://doi.org/10.1007/978-3-030-69322-0_27

4. Bellifemine, F.: FIPA: a standard for agent interoperability. In: WOA 2000: Dagli Oggetti agli Agenti. 1st AI*IA/TABOO Joint Workshop "From Objects to Agents": Evolutive Trends of Software Systems, 29–30 May 2000, Parma, Italy, p. 121. Pitagora Editrice Bologna (2000)

5. Bergenti, F., Caire, G., Monica, S., Poggi, A.: The first twenty years of agent-based software development with JADE. Auton. Agent. Multi-Agent Syst. **34**(2), 1–19 (2020). https://doi.org/10.1007/s10458-020-09460-z

6. Boissier, O., Bordini, R.H., Hübner, J.F., Ricci, A., Santi, A.: Multi-agent oriented programming with JaCaMo. Sci. Comput. Program. **78**(6), 747–761 (2013)

7. Chen, X., Song, H.: Further extensions of FIPA contract net protocol: threshold plus DoA. In: Haddad, H., Omicini, A., Wainwright, R.L., Liebrock, L.M. (eds.) Proceedings of the 2004 ACM Symposium on Applied Computing (SAC), Nicosia, Cyprus, 14–17 March 2004, pp. 45–51. ACM (2004). https://doi.org/10.1145/967900.967914

8. Ciortea, A., Boissier, O., Ricci, A.: Engineering world-wide multi-agent systems with hypermedia. In: Weyns, D., Mascardi, V., Ricci, A. (eds.) EMAS 2018. LNCS (LNAI), vol. 11375, pp. 285–301. Springer, Cham (2019). https://doi.org/10.1007/978-3-030-25693-7_15

9. Clemen, T., et al.: Multi-agent systems and digital twins for smarter cities. In: Giabbanelli, P.J. (ed.) SIGSIM-PADS 2021: SIGSIM Conference on Principles of Advanced Discrete Simulation, Virtual Event, USA, 31 May - 2 June, 2021, pp. 45–55. ACM (2021). https://doi.org/10.1145/3437959.3459254

10. Croatti, A., Gabellini, M., Montagna, S., Ricci, A.: On the integration of agents and digital twins in healthcare. J. Med. Syst. **44**(9), 161 (2020). https://doi.org/10.1007/s10916-020-01623-5

11. Eirinakis, P., et al.: Enhancing cognition for digital twins. In: 2020 IEEE International Conference on Engineering, Technology and Innovation (ICE/ITMC), pp. 1–7 (2020)

12. Fan, C., Zhang, C., Yahja, A., Mostafavi, A.: Disaster city digital twin: a vision for integrating artificial and human intelligence for disaster management. Int. J. Inf. Manag. **56**, 102049 (2021)

13. GE Digital: The digital twin: Compressing time to value for digital industrial companies. Technical Report, GE DIGITAL (2017). https://www.ge.com/digital/sites/default/files/download_assets/The-Digital-Twin_Compressing-Time-to-Value-for-Digital-Industrial-Companies.pdf

14. Gelernter, D.: Mirror Worlds or the Day Software Puts the Universe in a Shoebox: How Will It Happen and What It Will Mean. Oxford University Press Inc, New York (1991)

15. Glaessgen, E., Stargel, D.: The digital twin paradigm for future NASA and US air force vehicles. In: 53rd AIAA/ASME/ASCE/AHS/ASC Structures, Structural Dynamics and Materials Conference (2012)

16. Grieves, M., Vickers, J.: Digital twin: mitigating unpredictable, undesirable emergent behavior in complex systems. In: Kahlen, F.-J., Flumerfelt, S., Alves, A. (eds.) Transdisciplinary Perspectives on Complex Systems, pp. 85–113. Springer, Cham (2017). https://doi.org/10.1007/978-3-319-38756-7_4

17. Gutierrez, C., Sequeda, J.F.: Knowledge graphs. Commun. ACM **64**(3), 96–104 (2021)

18. Hribernik, K., Cabri, G., Mandreoli, F., Mentzas, G.: Autonomous, context-aware, adaptive digital twins - state of the art and roadmap. Comput. Ind. **133**, 103508 (2021). https://doi.org/10.1016/j.compind.2021.103508

19. Jennings, N.R.: An agent-based approach for building complex software systems. Commun. ACM **44**(4), 35–41 (2001)

20. Juarez, M.G., Botti, V.J., Giret, A.S.: Digital Twins: Review and Challenges. J. Comput. Inf. Sci. Eng. **21**(3) (2021). https://doi.org/10.1115/1.4050244

21. Kapteyn, M.G., Knezevic, D.J., Willcox, K.: Toward predictive digital twins via component-based reduced-order models and interpretable machine learning. https://doi.org/10.2514/6.2020-0418

22. Lippi, M., Mariani, S., Zambonelli, F.: Developing a "sense of agency" in IoT systems: preliminary experiments in a smart home scenario. In: 19th IEEE International Conference on Pervasive Computing and Communications Workshops and other Affiliated Events, PerCom Workshops 2021, Kassel, Germany, 22–26 March 2021, pp. 44–49. IEEE (2021). https://doi.org/10.1109/PerComWorkshops51409.2021.9431003

23. Liu, Y., et al.: A novel cloud-based framework for the elderly healthcare services using digital twin. IEEE Access **7**, 49088–49101 (2019)

24. Mariani, S., Omicini, A.: Anticipatory coordination in socio-technical knowledge-intensive environments: behavioural implicit communication in MoK. In: Gavanelli, M., Lamma, E., Riguzzi, F. (eds.) AI*IA 2015. LNCS (LNAI), vol. 9336, pp. 102–115. Springer, Cham (2015). https://doi.org/10.1007/978-3-319-24309-2_8

25. Minerva, R., Crespi, N.: Digital twins: Properties, software frameworks, and application scenarios. IT Professional **23**(1), 51–55 (2021). https://doi.org/10.1109/MITP.2020.2982896

26. Minerva, R., Lee, G.M., Crespi, N.: Digital twin in the IoT context: a survey on technical features, scenarios, and architectural models. Proc. IEEE **108**(10), 1785–1824 (2020)

27. Omicini, A., Ricci, A., Viroli, M.: Artifacts in the A&A meta-model for multi-agent systems. Auton. Agent. Multi-Agent Syst. **17**(3), 432–456 (2008)

28. Omicini, A., Ricci, A., Viroli, M., Castelfranchi, C., Tummolini, L.: Coordination artifacts: Environment-based coordination for intelligent agents. In: Proc. of the 3rd Int. Joint Conference on Autonomous Agents and Multiagent Systems, pp. 286–293. AAMAS 2004, IEEE Computer Society, USA (2004)

29. Orozco-Romero, A., Arias-Portela, C.Y., Saucedo, J.E.A.M.: The use of agent-based models boosted by digital twins in the supply chain: a literature review. In: Vasant, P., Zelinka, I., Weber, G.-W. (eds.) ICO 2019. AISC, vol. 1072, pp. 642–652. Springer, Cham (2020). https://doi.org/10.1007/978-3-030-33585-4_62

30. Papacharalampopoulos, A., Stavropoulos, P., Petrides, D.: Towards a digital twin for manufacturing processes: applicability on laser welding. Procedia CIRP **88**, 110–115 (2020). https://doi.org/10.1016/j.procir.2020.05.020

31. Rao, A.S., Georgeff, M.P.: Modeling rational agents within a BDI-architecture. In: Proceedings of the 2nd International Conference on Principles of Knowledge Representation and Reasoning (KR'91). Cambridge, MA, USA, 22–25 April 1991, pp. 473–484. Morgan Kaufmann (1991)

32. Ricci, A., Croatti, A., Mariani, S., Montagna, S., Picone, M.: Web of digital twins. ACM Trans. Internet Technol. (2021). https://doi.org/10.1145/3507909, Just Accepted
33. Ricci, A., Piunti, M., Tummolini, L., Castelfranchi, C.: The mirror world: preparing for mixed-reality living. IEEE Pervasive Comput. **14**(2), 60–63 (2015)
34. Saracco, R.: Digital twins: bridging physical space and cyberspace. Computer **52**(12), 58–64 (2019)
35. Shahat, E., Hyun, C.T., Yeom, C.: City digital twin potentials: a review and research agenda. Sustainability **13**(6) (2021)
36. Stary, C.: Digital twin generation: re-conceptualizing agent systems for behavior-centered cyber-physical system development. Sensors **21**(4), 1096 (2021)
37. Tao, F., Qi, Q.: Make more digital twins. Nature **573**(7775), 490–491 (2019)
38. Tao, F., Zhang, M., Nee, A.: Chapter 1 - background and concept of digital twin. In: Tao, F., Zhang, M., Nee, A. (eds.) Digital Twin Driven Smart Manufacturing, pp. 3–28. Academic Press (2019). https://doi.org/10.1016/B978-0-12-817630-6.00001-1
39. Uhlemann, T.H.J., Lehmann, C., Steinhilper, R.: The digital twin: realizing the cyber-physical production system for Industry 4.0. Procedia Cirp **61**, 335–340 (2017)
40. Valckenaers, P.: ARTI reference architecture – PROSA revisited. In: Borangiu, T., Trentesaux, D., Thomas, A., Cavalieri, S. (eds.) SOHOMA 2018. SCI, vol. 803, pp. 1–19. Springer, Cham (2019). https://doi.org/10.1007/978-3-030-03003-2_1
41. Wan, H., David, M., Derigent, W.: Modelling digital twins as a recursive multi-agent architecture: application to energy management of communicating materials. IFAC-PapersOnLine **54**(1), 880–885 (2021). https://doi.org/10.1016/j.ifacol.2021.08.104
42. Weyns, D., Omicini, A., Odell, J.J.: Environment as a first-class abstraction in multi-agent systems. Auton. Agent. Multi-Agent Syst. **14**(1), 5–30 (2007)
43. Ye, D., He, Q., Wang, Y., Yang, Y.: An agent-based integrated self-evolving service composition approach in networked environments. IEEE Trans. Serv. Comput. **12**(6), 880–895 (2019). https://doi.org/10.1109/TSC.2016.2631598
44. Zambonelli, F., Jennings, N.R., Wooldridge, M.J.: Developing multiagent systems: the gaia methodology. ACM Trans. Softw. Eng. Methodol. **12**(3), 317–370 (2003)

Only Those Who Can Obey Can Disobey: The Intentional Implications of Artificial Agent Disobedience

Thomas Arnold[1] , Gordon Briggs[2]([⊠]) , and Matthias Scheutz[1]

[1] Human-Robot Interaction Laboratory, Tufts University, Medford, MA 02155, USA
{thomas.arnold,matthias.scheutz}@tufts.edu
[2] Navy Center for Applied Research in Artificial Intelligence, U.S. Naval Research
Laboratory, Washington, DC 20375, USA
gordon.briggs@nrl.navy.mil

Abstract. Recent attention has been brought to robots that "disobey" or so-called "rebel" agents that might reject commands. However, any discussion of autonomous agents that "disobey" risks engaging in a potentially hazardous conflation of simply non-conforming behavior with true disobedience. The goal of this paper is to articulate a sense of what constitutes *desirable* and *true* disobedience from autonomous systems. To do this, we begin by discussing what it is not. First, we attempt to disentangle figurative uses of the term "disobedience" from those connotative of deeper senses of agency. We then situate *true* disobedience as being committed by an agent through an action that presupposes some understanding of the violated instruction or command.

Keywords: Machine ethics · Robot obedience/disobedience · Autonomy

1 Introduction

Robots are supposed to do what they are told. To many, the idea of a robot or artificially intelligent (AI) agent not doing exactly what is commanded by a human user raises alarming concerns: even examples of simple agents defying commands feature as a harbinger of doomsday scenarios of uncontrollable "super-intelligences" [4] or a robot uprising. While the likelihood and feasibility of these speculative scenarios are debatable, their undesirability is not. More concretely, there are various examples of current, real-world, autonomous systems acting in dangerous contradiction to well-accepted human regulations and norms. For instance, some current autonomous car prototypes have been reported as "disobeying" stop signs [22], an action that would likely lead to grave human harm if not corrected. Thus, whether the speculative robot apocalypse from science fiction or the more mundane autonomous car violating traffic laws, there is a clear space of behaviors exhibiting *undesirable* "disobedience", posing dangers that must be avoided.

© Springer Nature Switzerland AG 2022
F. S. Melo and F. Fang (Eds.): AAMAS 2022 Workshops, LNAI 13441, pp. 130–143, 2022.
https://doi.org/10.1007/978-3-031-20179-0_9

Are all instances of "disobedience" by autonomous agents undesirable? Recent attention has been brought to robots that "disobey" [6] or so-called "rebel" agents [7] that might reject commands [5] for good reasons – these may be desirable instances of "disobedience." Yet, the notion of "desirable disobedience" is not new. In his fiction, Isaac Asimov posited the three laws of robotics, which specify that a robot must obey orders given by humans (second law) except when they would lead to human harm (contra first law) [3]. Another example of robot "disobedience" in the service of avoiding human harm can be seen in the domain of "seeing eye" robots [17,23], where an assistive robot may steer its human handler away from hazards in an environment that the handler cannot perceive. The example of a "seeing eye" robot is an intuitive case where a human interaction partner may issue a command that is not aligned with his or her own overarching goals (i.e., safety). Even the most cautious skeptics of autonomous systems would likely concede that there is a space of agent behavior that could be considered *desirable* "disobedience."

Because robot and AI "disobedience" straddles the divide between desirable and undesirable behaviors, AI researchers have often approached the question of "disobedience" from the general standpoint of ensuring that any robot behaviors conform and/or are aligned with what humans consider desirable [21]. However, we contend that this approach is insufficient. In part, we argue that this is due to a potentially hazardous conflation of simply non-conforming behavior with true disobedience. Ask yourself: are all instances of robot or AI "disobedience" actually instances of disobeying something? Consider again the example of the Tesla autonomous car [22] that reportedly "disobeys" stop signs. It would seem incorrect to say that an autonomous car with computer vision algorithms that fail to correctly recognize a stop sign, or an autonomous car that has no knowledge of the difference between a complete and rolling stop, "disobeys" in a strong sense of the term. An autonomous agent with these sorts of limitations does not "disobey" in the same sense that a human-like agent would. To say that any AI or robot whose behavior does not conform to commands or norms "disobeys" may result in an over-ascription of agency and/or deliberative capability, an outcome not without its own unique set of dangers [20]. This slippage is especially important to correct given media incentives for clickbait and alarmist headlines, a propensity to magnify something like a sensor malfunction (e.g. one not picking up the lip of a doorway) as a robot breaking away for freedom [1].

The goal of this paper is thus to articulate a sense of what constitutes *desirable disobedience* from autonomous systems. To do this, we begin by discussing what it is not. First, we attempt to disentangle figurative uses of the term "disobedience" from those connotative of deeper senses of agency. We then situate *genuine* or *true* disobedience as being committed by an agent through an action that presupposes some understanding of the violated instruction or command. For interactions within instructional contexts, we argue that true disobedience (of the instruction) depends on a broader, more important form of obedience. For the sake of design clarity and future research, it is important to delineate what kind of "obedience" ascriptions disobedience relies upon, so that accountable and transparent systems can be more squarely pursued. A robot that says "no" for a reason it can express (and justify) and rely upon is one whose obedience and disobedience can be more accessible, correctable, and functional.

2   Senses of "Disobedience"

When we hear people say that something is "rebellious" or "disobedient", we can have a general understanding of what they mean. These descriptions can be applied to all manner of entities, ranging from pieces of equipment, animals, children, and adults. Yet, the implications we draw from hearing that someone's television is being "rebellious" are considerably different from the ones we draw from hearing that a person is being "rebellious". Likewise, "disobedience" has a similar range of contextually-driven connotations. Therefore, when we apply the terms "rebellion" or "disobedience" to an AI or other autonomous system, we must take care and be clear on what we precisely mean.

On one end, there is a sense of *figurative* "disobedience," in which the entity is simply behaving in a way that is unintended or undesired by a supervisory agent, but is correctly understood by the user or supervisor of being the sort of entity incapable of deciding to do otherwise. For example, we understand that a television that is failing to respond correctly to input commands from a remote control can be at best described as "disobedient" in a figurative way. On the other end, there is a sense of *true disobedience*, in which the entity is behaving in a way that is unintended or undesired by a supervisory agent, while also correctly understood by the supervisor as explicitly and deliberately deciding to do so. For example, a soldier refusing to obey an order given by a superior, citing potential violations of the rules of engagement or laws of war, would be understood as being disobedient in this truer sense.

However, this dichotomy still elides a significant amount of nuance. While obedience and disobedience may not require full agency in the sense of Moor [18], there are admittedly degrees of applicability for terms like "obedience" and "disobedience." For example, an unruly dog may not be disobedient in the same sense as a person, but we could say it is being disobedient in a sense more closely resembling true, rather than figurative, disobedience (though we will return to this point below). Furthermore, the cognitive complexity necessary for an agent to exhibit true disobedience would also likely entail a variety of contexts in which undesired behavior is not necessarily indicative of disobedience. For example, it would seem inappropriate to describe learning agents prior to or undergoing training as being disobedient. In other words, it would be less appropriate to describe an untrained dog that disregards a command as disobedient than to do so for a trained one that does the same. While a comprehensive treatment of these nuances is outside the scope of this paper, it is important to note that the degree to which the term "obedience" or "disobedience" can be applied to an autonomous agent's behavior in the non-figurative sense is directly tied with how the agent ostensibly produces these behaviors.

3   Recent Approaches to Robot "Disobedience"

Recently, researchers have used partially observable Markov decision processes (POMDPs) to formalize the notion of robot disobedience as corrective realignment of a specific human interactant's specific commands and larger aims [16]. In this work, Milli et al. (2017) investigate a trade-off between robot obedience and robot autonomy: "A blindly obedient [robot] R is limited by [a human] H's decision making ability. However, if R follows a type of [inverse reinforcement learning] IRL policy, then R is guaranteed a positive advantage when H is not rational." In other words, for the robot to be able to do better when a human is not rational, it needs to be disobedient.

While this seems to be an example of true disobedience at first glance, the relationship between autonomy and obedience (and disobedience) makes that murkier. For starters, systems that are trained to simply follow policies and/or learn rewards of their instruction givers from interactions with them (e.g., through IRL) will fail to appreciate norms instruction givers might be following. They can fail to register the difference these norms represent between the instruction giver's own preferences or those of the larger society (irrespective of the instruction giver's preferences). As a result, they might view an instructor as less rational and increase their propensity to not carry out that instructor's commands, but without a specific explanation of why a particular command was not followed. Without a represented choice, or a represented reason, or – most fundamentally – some rudimentary understanding that a command/instruction *is* a command/instruction, there is not disobedience but only nonconformity. And that lack of conformity can be described just as well as "dysfunction" if it violates enough global priorities/programming as it can be described as "disobedience". That is, only context-constrained reasons for refusals participate enough in the concept of "obedience" to be called disobedience in the true sense, rather than randomness, breakdown, and other events corresponding to figurative "disobedience." In sum, mere nonconformity in behavior does not mean something is being disobeyed.

Therefore, we would argue that robots that simply execute learned policies instead of weighing the applicability and tradeoffs among normative principles in a given context are neither obedient nor disobedient. Obedience and disobedience help define one another as ascriptions of an agent's capacities, not just external labels for behavior. Obedience thus requires the capability to disobey, which, in turn, requires an understanding of the possible norm violations implicated by particular actions, i.e., obedience and disobedience are relative to principles. A policy-based robot (regardless of how it learned the policy) is no different from a dish washer that starts the cleaning cycle when the right knob is turned. A norm-obeying robot, on the other hand, may have symbolically rendered knowledge, and (we will argue) *explicit* knowledge of normative principles that underwrite its actions. And if given a command, it can determine what principles are implicated and then make a decision based on those principles and the ones that might potentially be violated, to choose which principle to suspend and which to uphold (e.g., see [12] for such an approach).

It should also be noted that viewing robot obedience and disobedience as a dyadic problem, that is, considering the actions of a robot relative to the commands and aims of a single human user, has significant limitations. Many instances of disobedience are considered desirable not because a rejected command violates the issuer's true intent, but rather because the command contravenes larger legal or moral principles, regardless of the intent of the command-issuing agent. For example, ethical reasoning mechanisms have been proposed for hypothetical autonomous military systems to ensure that lethal force is deployed only when authorized by the rules of engagement and laws of war [2].

What this example indicates is a need to distinguish and detail the function of disobedience in contrast to the broader set of non-conforming or unexpected behavior on the part of an artificial agent. If there is a practical demand for a robot to explain its action, one that bears on other people acting in response, then disobedience is implicated in norms whose violation could threaten more than a robot's direct task. If there are norms or principles to which a robot's action is adhering, what are they? The ambiguity of divergent action, and the ascriptions it can incur, can blur important lines of accountability. What commitments does a robot's implicit or explicit "No" really make?

4   What Does True Disobedience Entail? Layers of Intentionality and Interaction

4.1   Understanding Instruction as Possible Action

Robots are used to perform tasks for which they have goals, implicit or explicit, regardless of how these goals made it into the system: through explicit instruction or the user's selection of one of multiple pre-defined options, or through learned policies for a given reward function. Performing a task then means for the robot to execute a sequence of actions that, if all goes well, will lead to the desired outcome. Actions taken by the robot will, in general, depend on the state of the environment, including the robot and other task-relevant objects and agents, and might thus change due to unforeseen events (e.g., the robot's effectors breaking) or actions by other agents (e.g., a human giving the robot a command that interferes with its goal). In such cases, the robot will attempt corrective actions to get back on track. In a policy-based system the best action under the current policy will be executed, while in a planning-based system the current plan will be abandoned and replanning will be triggered to determine the best course of action. Regardless of how a robot's response to circumstances that interfere with its task performance is initiated, changes in behavior to make progress towards the overall task goal per se have nothing to do with disobedience, even when these changes are the robot's reaction to a human instruction (e.g., the robot taking time to parse the speech that interrupts it).

Understanding an instruction functionally, wherein a system identifies it as a directive, is a level of intentionality that sets disobedience apart from merely failing to meet an instructed constraint. A sensory lack or a clumsy execution may prevent the system from following an instruction, but that is not at all a disobedient result. The oblivious or incompetent system merely fails at what it would do, instead of choosing failure itself.

4.2 Capability of Obedience

Thinking of robots in terms of obedience can heighten their association with animals (especially service animals or pets). "Obedience school", especially for dogs, is a common practice for getting one's pet to behave in conformity with instruction. It is worth noting carefully, however, that in that context the opposite of an obedient animal is not a disobedient one. It is an untrained one. If it does not learn some established connection between an explicit command and the expected action it is impervious to instruction: not just untrained but perhaps untrainable. This distinction, to the degree it matches one's intuitions, reflects the dependency that disobedience has on the capability to obey. That capability is more firmly assumed if there is proven performance thereof, if the agent has been discerning and reliable in obeying instructions before. If there are two dogs being commanded to sit while a bunny runs into each of their fields of vision, it is the trained one – the one that knows to assume the position that satisfies "Sit!" and under ordinary circumstances does so without hesitation – that one would ostensibly call disobedient by running away. The other dog, one that has never heard "Sit!" before, is not disobeying anything. There is nothing in its experience or learned capabilities to defy.

In child psychology, the dependence of disobedience on obedience takes on even more intentional dimensions. Seen in both everyday instruction and scholarly research, the lines between intentional and accidental action are continually negotiated as children develop the ability to interpret the intentions of other agents as well as classify, if not justify, their own actions [8]. Actions that are "unexpected" or "wrong" can be accidents, and children learn to apply degrees of intentionality before applying blame to agents, what they "knew full well" before doing [11]. The ability to obey, an established understanding and execution of its goal, informs when an agent "knew" what it was doing relative to what was expected or demanded. The ongoing moral education of children will refine differences between justified reasons and bad excuses, but throughout there will be intentional ascriptions of what the agent "knew enough" to do (hence knew enough to be held responsible for doing wrong). As one considers artificial agents and obedience, in fact, it is suggestive to look at how children interpret deviant or unexpected robotic action. Lemaignan et al. [14] show that what they set up as "disobedient" behavior does not often get interpreted as such by their child participants, to the degree there is less ascription of intentionality to the robots – they already have a sense that such a behavior is not disobedience unless there is some intentionality that defines the action.

The capability of obedience is part of the implicit intentionality connecting instruction to an agent. If an agent has never obeyed a specific instruction, nor shown a response that reflects an inbuilt relationship between instruction and executable action, on what basis would one relate its actions to the instruction at all? The perceptual proximity of an instruction does not anchor it *qua* instruction toward the agent. I may yell at a dolphin leaping close to me in the waves to come give me a ride to Australia, but what would that have to do with defining its ensuing behavior?

4.3   Reason, Purpose, or Commitment for Acting Against Instruction

The capability of obedience in a system establishes that a system is matching its action with the content of a command, not just in coincidental conformity with a command. The difference hangs on how the instruction brings about that action based on the system's design. An non-conforming action lacking any intentional relation to an instruction may be indistinguishable behaviorally from one emerging from an instruction being outweighed, overridden, or rejected on some set of terms. Disobedience, by extension, does not represent a pure severance from instruction but a more complicated relationship with it. Disobedience is not a inability to take instruction and obey. On the contrary, it is a conditional rejection of instruction, one that invites a query into what conditions explain such rejection and subsequent disobedience.

One can consider as an illustration two policy-based systems that are seeking to maximize a reward in each state they find themselves occupying. An action from each that conflicts with an instruction may have two different explanations – in the one case the system knows certain features of the state space better than the instructor, and is selecting better actions based on that knowledge. Its action just happens to conflict with an instruction that it is not incorporating into its decision. The other system simply has an estimate of the instructor's rationality that justifies partial divergence from what they instruct. Each system can be seeking to maximize reward, but they start from different points of training. If the latter system learns more about the state space than the former, it will converge in action and have its own training to depend upon – the instructor need have no further effect on the process. Disobedience need not apply to either system on an intentional level, though divergent actions from each are still observed.

In a planning system with explicit reasons (or purposes, or commitments), however, disobedience retains its sense as the deliberate rejection of an obedient action. There is a condition that meets a standard for acting against instruction, a reason that justifies taking the disobedient course.

Without such layers of reason, purpose, and commitment, one is left with other ways that work better to describe a system: errant, untrained, impervious, oblivious, malfunctioning. These are various ways of describing systems for which the instruction is not a represented object of planning or decision-making. Or there is no capability established of obeying what is represented. Or divergence from instruction has no intentional basis to it. In order to lay out how disobedience on the part of a robot would or should work, therefore, it is important to delineate 1) what such an order or guide is, 2) what design feature the system possesses to incorporate that instruction. If there is no such identifiable order in a given situation or environment, then the robot's disobedience is only a hypothetical attribution, performing *as if* it were disobeying some order that the observer infers or makes up to put context for the robot's action. If the system has no design feature or architecture by which an order can affect its operation, then its violations or conformity toward rules out in the world are, at best, inadvertent.

The more one loosens intentions from the constitution of the action, the less disobedience can be said to differ from incidental divergence. An agent that is choosing an action for the sake of the highest reward may take the exact same action as one that is disregarding an instruction from an irrational instructor. How would one tell the difference, and when would it matter to know the difference?

One can also view the reason for disobedience as part of an implied set of counterfactuals, what needs to be the case for an instruction to be obeyed or disobeyed. If there are no conditions under which a system could both understand and competently follow an instruction, yet still not decide to do so, then one might ask if obedience and disobedience still apply. While this paper cannot explore this implication adequately, it is worth considering what degree of weighing or judging competing reasons for following an instruction is implied by obedience and disobedience both.

5   Local vs. Global Disobedience

One way to think about conflicting forms of obedience is through competing principles. Some ethical dilemmas can arise when two explicit rules cannot both be upheld, and one must account for why, and to what effect, one chooses to disobey one rather than the other. For this paper, we propose that cases of "rebel" or "disobedient" agents can be thought of as representing more or less local, and more or less global, norms and priorities. Not only do there need to be choices made as to which norm or guideline takes precedence over another, but one must ask where that precedence comes from and what enforces it. The question this embeds in matters of AI system design is whether "obedience" and "disobedience" even apply to a system designed and implemented without reference to these levels.

The better alternative to using concepts of obedience, for systems that have no internal reference to norms or ability to interact with explicit reference thereto, is to say they conform or diverge (and perhaps functions or malfunction). The intentionality at work is about what the system is "supposed" to do by the designers, without any internal deliberation about that which could be obeyed or disobeyed.

Let us consider an example of different robotic systems implemented within a hospital. The demands and challenges of COVID have only heightened questions of how robots might be useful in such a context, since they could help keep an environment sterile while performing basic tasks [10]. Imagine a delivery robot going down hallways carrying supplies from one part of the hospital to another. It has a limited natural language repertoire related to moving, stopping, and alerting others to its intended task. Let us also suppose there is a robotic system that serves as an informational kiosk, largely staying put and responding to questions from visitors.

There is no necessity of either robot being disobedient. Each robot might only respond to queries or instructions that fall within its assigned task, and lack a capability to understand or obey anything else. Perhaps by certain design elements of appearance (e.g., screen vs. no screen) visitors would not confuse the two kinds. This, again, would be robots that were not able to obey certain instructions, meaning they do not disobey them either.

Now consider the delivery robot operating during a security emergency, when hospital policy is to have all mobile robots cease operating while it is addressed. Its stopping could be interpreted in different ways by hospital staff. Perhaps its navigation is malfunctioning, perhaps its effectors are. But if it is instructed "Take these to Room 206", it would stop from functional disobedience because of a security protocol. "This delivery cannot proceed for security" would be the more global directive that would justify its local disobedience.

Here it may be worth recalling Mirsky and Stone's "seeing-eye robot," which could disobey unsafe commands from a user who does not detect a harm that the robot perceives [17]. This would mark a difference in access to information about the world. For situations like that of a delivery robot's situation, however, there might be shared knowledge about the protocol: the important point is that the robot uphold the right priority, whether it provides new information or not.

Is there a global form of disobedience for a system, where every designed standard or instruction is rejected? If there is, much less if it were sought after, then one should ask how this differs from malfunctioning, and harmful malfunctioning at that. What design reasons or purposes would such disobedience serve, if not some larger aim or principle? And if there is no question of deciding between obedience and disobedience, because there is no distinct process attached to receiving an instruction to obey, then the concept functionally falls out of the description.

Distinguishing local and global forms of disobedience allows one to compare an agent's particular actions relative to an instruction and a background norm system, allowing for modal descriptions of obedience and disobedience.. Take two norm systems A and B, where an instruction I violates A but not B. A robot trained to uphold A performs as designed if it disobeys I, whereas the robot trained to uphold B may not be obligated to follow I in order to uphold B. Robot B may disobey for local consideration (a near-simultaneous, but incompatible, instruction followed first), whereas Robot A disobeys for a global one.

6   Transparency and Accountability for Ascriptions of Disobedience

As the introduction of self-driving cars heated up several years ago, more attention was given to how, as an autonomous system, a car should obey its owner/driver's own values and priorities. The notion of "moral proxy" is one way to describe what a locally disobedient, globally obedient system could represent, acting on behalf of an agent or community who was not directly instructed the robot in the moment [15]. The system itself, by opting to override an immediate instruction for the sake of an overarching norm, is not by virtue of that a "full ethical agent" [18]. Its upholding of that norm is more plausibly seen as a proxy for some community or societal decision about how vehicles should operate, regardless of whether a particular owner wants. Alternatively, a system that disobeys the larger societal rule for the sake of an owner's instruction would be disobedient as the owner's moral proxy. Distinguishing levels of obedience goes hand in hand with locating the moral proxy at work.

There are, of course, plenty of ways that the system could evoke overly robust ascriptions of agency, incurring blame for its action. The manufacturers, government regulators, owners, and others could claim some malfunction was the reason for the conflict, distancing themselves from difficult decisions amid norm conflicts by branding it "going rogue". Media and entertainment harp on the them of creations turning on their creators, as many texts, from Genesis to Mary Shelley's Frankenstein, have helped summon. But these cloud the harder work of deciding what norms a robot should uphold, how design ought to achieve true accountability to them, to which norms a robot should be ultimately obedient. While liability of technical failures will always be a thorny issue to settle, designing and implementing systems without norm transparency is a serious social and technical risk. No romance of heroic rebellion or opaque algorithmic insight should obscure the importance of norms in the social fabric, as well as the ordinary demands for accountability and reason-giving that helps that fabric hang together.

The concept of disobedience implicates a role for instruction, a consideration of that instruction, and a justifiable decision to act against instruction. There are terms other than "disobedient" to describe more precisely how agents do not behave in conformity with an instruction. If an instruction is not something a system can understand or integrate into its operation – if it makes no difference to how the system chooses its action – then it is better to call it "impervious to instruction". If a system merely diverges from expected or emergent patterns (e.g., emergent coordination in a multi-agent simulation), without any explicit representation or expression of why, then it should be called "errant" (in a neutral sense of taking unexpected paths). Many impervious or errant agents could look like "disobedient rebels"; but without some feature that operationalizes an order or instruction, obedience and disobedience do not apply to them.

6.1 Obeying an Instructor vs. an Instruction

One response to the argument for restricting disobedience to more intentional forms is to say that, for an RL or IRL system, there is obedience and disobedience of an instructor, not so much instruction [9]. While there might not be explicit representations, much less natural language expressions, of what is being asked of an agent, it still makes a certain amount of sense to say "Obeying or disobeying the agent just means following or not following what the instructor commands at time t". The estimation of disobedience would, then, be how to optimize a policy with an instructor whose directions might not be fully rational or accurate.

Within the confines of a dyadic relationship exploring a simple state space, this has some cogency. For more social and symbolic interactions among other agents, however, or even reasoning across time between two agents about priorities [13], the application of disobedience loses sense. Are there any rules, norms, or orderings that other agents could understand as objects of obedience and disobedience between the original two? What is the content of the disobedience being interpreted by others who would coordinate action with the artificial agent, if that agent is to be thought of as taking purposeful action rather than malfunctioning or coming up short in its execution? If the idea is that learning reward functions from an environment may mean divergent exploring, it might make sense to call such conforming or non-conforming actions "ignoring a suggested action" and "following a suggested action". There is no independent rule being reasoned over or possibly shared with other agents, there are no explicit inferences or beliefs that can cited as reasons for its actions. In sum, the model of obeying an instructor alone is so thin a conception of sociality and rationality as to render "disobedience" without clear validity or purpose.

7   What Kinds of Disobedience Should Be Sought?

Going forward it is crucial to distinguish obedience and disobedience as localized, if rudimentary, operations on the part of an interactive system from obedience and disobedience as *ascriptions* from human interactants. For systems that, as we have argued, can neither obey nor disobey by design, one could still anticipate ascriptions of obedience and disobedience by those unfamiliar with its lack of capacity. In that case one might speak of ascription mitigation, the avoidance of implementation that evokes a deluded sense that a "rebel agent" is on the loose. But that is, to repeat, quite different from actual obedience and disobedience to a concrete guideline, through a system's designed inference that a larger norm or rule is to be followed. This distinction should guide future research lest forms of "rebellion" obscure deliberate decisions as to what kind of policies, symbolic representations, and logical operators (or lack thereof) are behind the system's performance.

For the field of human-robot interaction, at least, it is worth investigating how ascriptions of disobedience detract or enhance social robotic applications. Are there more functional affordances that social robots could give, more transparent measures indicating what the system recognizes and reasons about as instruction, to deter or guide more instinctive judgments toward their actions [19]?

The point of this paper is not to sequester the term "disobedience" behind the loftiest of intentional standards. The ascription of intentionality can vary with contexts, and in some multi-agent environments it may serve as a useful heuristic to call certain systems "disobedient" so that other people do not expect instruction to be effective. But for more general discussions of accountability and design, as well as public-facing discussions of robots who elicit the label of incipient "overlords", it is important to take more care about what disobedient is really saying about the system.

The intentional invitation that "disobedience" makes to various discussions around AI and robotics is difficult to disentangle from the common sense judgments and practical reasoning that communities and societies employ. While there is no clean way to prevent the overuse of intentional terms toward artificial agents, that does not mean the invitation should go unattended and unrestrained in robotic applications. To avoid exploitative and manipulative uses of intentional language, especially a morally charged term like "disobedience", public discussions ought to reflect clarity and carefulness in technical offerings. A disobedience devoid of intentionality and without systemic transparency risks turning the intentional invitation into a provocation to fear, fantasy, and hype.

The twofold task around disobedience, then, is 1) to state what dimension of intentionality defines a system's ability to act upon a command, principle, or rule, to qualify how "disobedience" ought to be applied, 2) to map what kind of disobedience, with what form of intentionality, can most responsibly feature in a system's real-world interactions and applications.

Public discussion of robots, fueled in part by depictions of robots in film and television, often evokes the threat of rebellion or takeover from any seeming independence on the robot's part. The line "I'm sorry, Dave, I'm afraid I can't do that" from 2001: A Space Odyssey is infamous for the terror caused by a machine untethered by human command. Our discussion of disobedience, including the framework of a local vs. global disobedience, is to push against this cluster of associations. If disobedience is intentionally possible for a robot, then the difficult questions go to its design and how its intentions conform to the priorities of those affected by it. If it is not truly disobedient, then the responsible question is to what degree it can take instruction – its supposed obedience and disobedience may be illusory, and its disconnection from human instruction even more recklessly severe.

8   Conclusion

The idea of disobedience from artificial agents carries two important points for design. First, disobedience is a term with intentional implications and connotations, and ignoring these can easily misrepresent what the system is doing and how it is to be held accountable. A system that disobeys is one that is equipped to obey, capable of obeying, but has an accessible reason to take an alternative course of action. Second, for a system to disobey responsibly, the reasons to obey or disobey must be specified and ordered in an accountable fashion. The priorities, norms, and commitments that take precedence over others, however rudimentary they are in context, should define the need for and function of disobedience. Impervious or errant systems may learn to navigate the world, but their lack of conformity is not an intentional disobedience; consequently, the design of a properly and usefully *disobedient* robot must integrate and offer accurate access to reasons. Exaggerating and overpromising intentionality does not just fuel "rogue robots" hype on the way to public declamations of panic, it fundamentally attacks an interactive norm of transparency and accountability. If robots are to disobey for reasons that matter, reasons must matter in their decisions.

Acknowledgments. The views expressed in this paper are solely those of the authors and should not be taken to reflect any official policy or position of the United States Government or the Department of Defense.

References

1. The runaway robot: how one smart vacuum cleaner made a break for freedom (2022). https://www.theguardian.com/lifeandstyle/2022/jan/24/the-runaway-robot-how-one-smart-vacuum-cleaner-made-a-break-for-freedom
2. Arkin, R.: Governing Lethal Behavior in Autonomous Robots. Chapman and Hall/CRC (2009)
3. Asimov, I.: Runaround. Astounding Sci. Fiction **29**(1), 94–103 (1942)
4. Bostrom, N.: Superintelligence. Paths, Dangers, Strategies (2014)
5. Briggs, G., Scheutz, M.: "Sorry, I can't do that": developing mechanisms to appropriately reject directives in human-robot interactions. In: AAAI Fall Symposium Series: Artificial Intelligence and Human-Robot Interaction (2015)
6. Briggs, G., Scheutz, M.: The case for robot disobedience. Sci. Am. **316**(1), 44–47 (2017)
7. Coman, A., Aha, D.W.: AI rebel agents. AI Mag. **39**(3), 16–26 (2018)
8. Flavell, J.H.: The development of children's knowledge about the mind: from cognitive connections to mental representations (1988)
9. Hadfield-Menell, D., Russell, S.J., Abbeel, P., Dragan, A.: Cooperative inverse reinforcement learning. Adv. Neural Inf. Process. Syst. **29** (2016)
10. Holland, J., et al.: Service robots in the healthcare sector. Robotics **10**(1), 47 (2021)
11. Kalish, C.W., Cornelius, R.: What is to be done? children's ascriptions of conventional obligations. Child Dev. **78**(3), 859–878 (2007)
12. Kasenberg, D., Scheutz, M.: Interpretable apprenticeship learning with temporal logic specifications. In: 2017 IEEE 56th Annual Conference on Decision and Control (CDC), pp. 4914–4921. IEEE (2017)
13. Kasenberg, D., Scheutz, M.: Norm conflict resolution in stochastic domains. In: Proceedings of the AAAI Conference on Artificial Intelligence, vol. 32 (2018)
14. Lemaignan, S., Fink, J., Mondada, F., Dillenbourg, P.: You're doing it wrong! studying unexpected behaviors in child-robot interaction. In: ICSR 2015. LNCS (LNAI), vol. 9388, pp. 390–400. Springer, Cham (2015). https://doi.org/10.1007/978-3-319-25554-5_39
15. Millar, J.: Technology as moral proxy: autonomy and paternalism by design. IEEE Technol. Soc. Mag. **34**(2), 47–55 (2015)
16. Milli, S., Hadfield-Menell, D., Dragan, A., Russell, S.: Should robots be obedient? In: Proceedings of the 26th International Joint Conference on Artificial Intelligence, pp. 4754–4760 (2017)
17. Mirsky, R., Stone, P.: The seeing-eye robot grand challenge: rethinking automated care. In: Proceedings of the 20th International Conference on Autonomous Agents and Multi-Agent Systems, pp. 28–33 (2021)
18. Moor, J.: Four kinds of ethical robots. Philos. Now **72**, 12–14 (2009)

19. Paauwe, R.A., Hoorn, J.F., Konijn, E.A., Keyson, D.V.: Designing robot embodiments for social interaction: affordances topple realism and aesthetics. Int. J. Soc. Robot. **7**(5), 697–708 (2015)
20. Robinette, P., Li, W., Allen, R., Howard, A.M., Wagner, A.R.: Overtrust of robots in emergency evacuation scenarios. In: 2016 11th ACM/IEEE International Conference on Human-robot Interaction (HRI), pp. 101–108. IEEE (2016)
21. Russell, S., Dewey, D., Tegmark, M.: Research priorities for robust and beneficial artificial intelligence. AI Mag. **36**(4), 105–114 (2015)
22. Sparkes, M.: Tesla recalls 50,000 cars that disobey stop signs in self-driving mode. New Scientist (2022). https://www.newscientist.com/article/2307147-tesla-recalls-50000-cars-that-disobey-stop-signs-in-self-driving-mode
23. Tachi, S., Komoriya, K.: Guide dog robot. Auton. Mob. Robots: Control Planning Archit. 360–367 (1984)

Correction to: Data-Driven Agent-Based Model Development to Support Human-Centric Transit-Oriented Design

Liu Yang and Koen H. van Dam

Correction to:
Chapter 3 in: F. S. Melo and F. Fang (Eds.): *Autonomous Agents and Multiagent Systems. Best and Visionary Papers*, LNAI 13441, https://doi.org/10.1007/978-3-031-20179-0_3

In the originally published version of chapter 3, the funding information in the acknowledgement section was incomplete. This has been corrected.

The updated version of this chapter can be found at
https://doi.org/10.1007/978-3-031-20179-0_3

© Springer Nature Switzerland AG 2024
F. S. Melo and F. Fang (Eds.): AAMAS 2022 Workshops, LNAI 13441, p. C1, 2024.
https://doi.org/10.1007/978-3-031-20179-0_10

Author Index

Printed in the United States
by Baker & Taylor Publisher Services

Printed in the United States
by Baker & Taylor Publisher Services